Handbook of RELOADING BASICS

Robin Sharpless and Rick Sapp

Published by

Gun Digest® Books, an imprint of F+W Media, Inc.
Krause Publications ▪ 700 East State Street ▪ Iola, WI 54990-0001
715-445-2214 ▪ 888-457-2873
www.krausebooks.com

To order books or other products call toll-free 1-800-258-0929
or visit us online at www.gundigeststore.com

ISBN-13: 978-1-4402-4132-1
ISBN-10: 1-4402-4132-5

Cover Design by Kevin Ulrich
Designed by Sandi Carpenter
Edited by Jennifer L.S. Pearsall

Printed in the United States of America

About the Authors

Robin Sharpless currently serves as the Executive Vice President of Redding Reloading Equipment. His career in the firearms industry now spans three decades, with senior level positions in both the civilian and military markets. During this time, he has served on multiple SAAMI and NRA committees, lobbied Congress with the American Shooting Sports Council, and has formerly served on the NRA Board of Directors. His company tenures have included executive positions with H&R 1871, Ithaca Gun, and Cheytac, and he is a recognized reloading subject matter expert (SME) for the International Association of Firearms and Tool Mark Examiners.

Author Rick Sapp is a popular writer of many outdoor subject matters, with noted expertise in shotgunning. He has previously authored several volumes for Gun Digest, including *Standard Catalog of Colt, Gun Digest Book of Trap and Skeet, The Gun Digest Book of Green Shooting: A Practical Guide to Non-Toxic Hunting and Recreation*, and *Reloading for Shotgunners,* from which much of this text was derived and updated.

Introduction

The practice of reloading ammunition has taken on explosive, almost unimaginable popularity in recent times. The reasons for this are several. The current political climate and the push for ever-tightening firearm and ammunition regulations has gun owners on the defensive. The mindset of "buy it before they take it away" has become commonplace among shooters of all walks, from weekend recreational and one-gun self-defense practitioners to serious competition shooters and avid hunters. The current state of the firearms industry is also spawning new shooters every day, with those who either didn't have a position on gun ownership before or who were on the fence now opening their minds and wallets and exploring all that the shooting sports has to offer.

In line with record-breaking gun sales across the country, ammunition—or the lack thereof—has been the hot topic for much of the last couple years. Retail stores were sold out for months at a time of everything from .22 LR to .30-06 to shotshells of all sizes, with reports of hoarding by many. Some sellers placed limits on ammunition buys—two boxes, five boxes—hoping to expand their customer base. At the same time, factory ammunition makers went into full swing, pumping out fresh rounds as fast as possible, but when store shelves remained empty for legions of shooter, they swarmed to reloading.

The numbers of consumers new to reloading is staggering—and exciting! And that's why we've assembled this book. We've all been in those shoes. You're faced with a complicated set of tools, tools that often don't look like they'll do what they do, and that leaves many new to the practice intimidated. Which type of press is best? How do I make it work once I have it? What components do I need to put my ammo together, and will my ammo look as good as the factory? These and other questions can stop the new reloader before they start.

That's why we've put together *Handbook of Reloading Basics*. Intended not just to get you started, but also to become a volume you keep as a practical and handy reference on your reloading bench or in your range bag, authors Robin Sharpless and Rick Sapp cover all those first questions the new reloader has and more. Indeed, we've covered both metallic cartridge and shotshell reloading, with advice on press and accessory buys, component necessities, and step by step instructions for your first time at the press. You'll also find a reference section full of reloading and component suppliers, and these are the experts you'll turn to for advice on specific powders, extra tools that streamline the reloading process, and the specialty equipment that can take your accuracy and shooting to new levels.

Whether you're a homeowner who keeps just a simple 9mm handgun by the beside for home-defense, a hunter who enjoys the challenge of extending your range, or a weekend skeet or sporting clays shooter who wants to start competing and sees the value in saving money by reloading, *Handbook of Reloading Basics* has something for everyone. Enjoy—you've just embarked on one of the most interesting aspects of shooting.

Table of Contents

PART I:

SECTION A: Metallic Cartridge Reloading Fundamentals

1 What's It All About?

Metallic cartridge reloading is a marriage of the process and techniques used to recycle a previously fired brass cartridge case for reuse as a newly loaded round of ammunition. The goal is to produce a round consistent with the original form and function of the factory round from which it was derived. The process prepares the case for reuse by cleaning, resizing to a factory specification, and finishing with the replacement of the components that have been either consumed or discharged from the firearm in the firing process. The terms "reloading" and "handloading" are often used interchangeably, but reloading truly differs from the conceptual process of handloading, which is performed to produce a "custom" cartridge configuration generally not available in a factory-supplied cartridge. Factory ammunition, for the purposes of this discussion, is that produced in the U.S. by members of SAAMI, the Sporting Arms and Ammunition Manufacturer's Association. SAAMI is the governing body of U.S. firearms and

Reloaded cartridges made to SAAMI specs.

Reloading today is now thankfully beyond this massive "portable" shotshell reloading setup (seen here back in the day at the Hornady plant), and the handloading of blackpowder firearms. We've come a long way, baby!

ammunition makers and sets the standards adopted and used by ANSI, the American National Standards Institute.

In this text, we'll discuss the basic principles and specific steps required to reload metallic cartridge cases in both bottleneck and straight-wall cases. Generally, these two are considered as rifle and pistol cases, respectively, but there are straight-wall rifle rounds and bottleneck pistol or revolver cartridges (an exception to every rule). In the end, you will have a basic understanding of the needed concepts and safety protocols to build good quality ammunition for use in your centerfire metallic cartridge firearms at a fraction of the cost of buying newly produced factory ammunition.

This first section of *Handbook of Reloading Basics* strives to give the beginning reloader or the individual considering getting into the hobby of reloading a fundamental understanding of the tools, processes, techniques, and safety practices necessary to produce accurate, safe, and reliable ammunition for both rifles and handguns. This book is highly illustrative in nature, allowing for a visual understanding of reloading that's supported by the written component. It strives to present a complete but basic understanding of this exciting, enjoyable, and money-saving addition to the sport shooter's hobby.

2 Metallic Cartridge Components

There are four parts to the basic centerfire metallic cartridge case, whether handgun or rifle (rimfire cases are generally considered to be non-reloadable, as the reloading practices for such cartridges require advanced ballistics knowledge and component handling). The parts to a centerfire metallic cartridge case are the case (usually brass, but sometimes other materials), the primer, the powder, and the bullet. I'm going to walk you through each of these components in more detail in a bit, but before we jump right in, I think it's a good idea to take a step back and look at what goes on in a loaded cartridge when your gun's firing pin strikes the primer.

When we pull the trigger, a spring in your gun—this can be a flat spring, a "V" spring, or a coil spring, doesn't matter which, all you need to know for the purposes of this discussion is that the spring provides the energy needed to fire your loaded cartridge—powers a hammer or a striker to impact the rear of your gun's firing pin, whose sole purpose is to put a dent in the center of the primer, which is housed in the center of the base of your brass case. The dent, coming from the firing pin striking the metal side of the primer, creates the pressure needed to ignite the priming mixture. This creates a jet of flame that travels around the anvil and moves forward through a central "flash hole" at the bottom of the primer cup (contained inside the cartridge case head) and begins the process of igniting the powder charge.

Now let's look at what's happening at the other end, the bullet end. Here, the neck walls of the case and the crimp around the bullet (if one is present), hold the bullet in the case, allowing pressure to build from the primer and powder ignitions until a "start pressure" is attained. This is the amount of pressure necessary for the bullet to overcome the forces of the friction between the case neck walls (and possibly crimp) and the bullet they hold. Once this happens, the bullet begins its travel.

Start pressure is important in the firing process, as it sets the stage for all the pressure-related events to follow. Consistency of velocity of the bullet in flight after it leaves the barrel, as well as the minimization of the bullet's vertical dispersion downrange, all start with a consistent "bullet pull," or start pressure, if all other factors are equal.

The start pressure is the beginning of the "pressure curve." The pressure curve is actually two bumps in succession. The initial pressure curve is set up by the bullet pulling out of the case. As the bullet begins to move, pressure in the case behind it drops, as the volume of area within the case is increased by the bullet's forward motion. As the bullet moves through the forcing cone, or "lead," (pronounced "leed") in the barrel, pressure continues to drop. (Bullet seating depth is important at this stage, so that the pressure drop is not too great, and this is something I'll discuss later.)

Now the bullet impacts the rifling, which is smaller in diameter than the bullet. When the bullet reaches this stage in its travel, for the briefest of moments the bullet stops. This is the point that begins the second, major pressure curve, derived from a point predetermined by the first. At this time, the powder really begins to burn at a frantic rate.

Smokeless powders consume themselves at a rate based on the inherent heat and pressure *of the moment*. So let's look, in a slightly different manner, at what's going on when the bullet hits the lead. At that moment, the bullet has stopped and plugged the barrel. The powder now burns violently, creating the gasses and pressure necessary to push the bullet into the rifling, swaging that bullet with the lands and cutting the imprint of the rifling into the projectile as is passes down the barrel. This action is needed to spin the projectile on its axis, stabilizing it on its flight after it leaves the barrel.

The bullet now transits the length of the barrel and is expelled ahead of a large plume of the expanding gasses that once powered it. The bullet continues to make its way downrange, and we are left with an empty cartridge case.

Let's review and think about what was used, lost, and consumed during the process:

1. Primer—impacted, fired, and spent.
2. Primer Cup—remains in the cartridge case and must be removed during the initial stages of the reloading process.
3. Powder—consumed and turned to hot gas and energy to propel the bullet.
4. Bullet—sent downrange never to be seen again (and, if it were found,

it would be unusable again, due to rifling imprints and deformation resulting from the final impact).

With these facts in mind, now we can look at each of the four components in more detail.

CASES

Metallic cartridge cases for your reloading will generally come from previously fired rounds of factory ammunition (known as once-fired brass). The best of these will be the case you previously shot in your own distinct firearm. Every firearm is slightly different in manufacture, with tiny but important differences in chamber, lead, bore, and rifling dimensions, each of these dimensional differences has an impact on the case during firing and so each must be accounted for in the reloading process. Therefore, cases that you obtain that have first been fired in your own gun will, therefore, will be better tailored to reloads meant to function in that same firearm. But let's say you don't have once-fired brass from your gun. Let's say you've purchased a batch of brass in the correct caliber from an online resource. That's perfectly okay to do. Cartridge cases have a set of SAAMI—Sporting Arms and Ammunition Manufacturers' Institute; www.saami.org—specifications governing their dimensions, and these SAAMI specifications have a group of minimum/maximum tolerances to which manufacturers adhere. Manufacturers are free to work within these bounds. Of course, you'll have to consider those minor variations, when engaging in the reloading process.

If you are using "range brass," those pieces of brass in the caliber you want that you've picked up after a session on the range, or brass that has been fired in a firearm other than your own but of like caliber and given to you, inspection and sorting becomes a far more detailed process. With such brass—brass that is the same caliber but not originally fired from your own gun—you need to first be sure the cases are truly for the correct chambering! A .270 Winchester and a .280 Remington can look the same, at first glance. *Always* check the headstamp for caliber designation as a first (and final) act of safety.

Another source of brass commonly used by reloaders is once-fired military brass, but there is a catch. Military brass cases in calibers like 5.56x45mm and 7.62x51mm are not exactly the same as .223 Rem. and .308 Win. (respectively). Differences in operating pressures necessitate thicker case walls in the military brass, making the internal space in those cases smaller, i.e., they have a lower case volume. If you used the

The markings on the bottom of your metallic cartridge case are known as the "headstamp." They generally (emphasis on "generally") tell you the cartridge name and, occasionally, the manufacturer of the original cartridge.

same powder charge in a military case of, say, 5.56, for example, that you used in a commercial .223 Rem., the lower volume in the 5.56 case results in higher pressures. This doesn't mean you can't use once-fired military brass, but you should sort these into their own bin and adjust a reloading recipe accordingly and specifically for that brass only, for the sake of safety. Also, if you decide to use military brass, doing so generally requires that you start with lighter loads than you would with their non-military counterparts (5.56 vs. .223 brass, as an example), in order to minimize the chances of high pressure issues. Truly, though, when starting any new loading regime and new cases, you should always begin with the lightest load listed in the data manual.

As a side note, you should know that military brass cases generally have "crimped–in" primers. Crimped primers are somewhat more difficult to remove; a dedicated "decapping" die is a benefit here, as it can handle this extra task with ease. Too, after removing the primer from such military brass, the remaining crimp ring must be removed from the primer pocket before reloading; without this step, a new primer

This is the thing that holds the rest of your load together. Generally constructed of brass, metallic cartridge cases will sometimes be found made of other materials.

will become damaged in the seating process. I like decapping tools that swage the brass of the case itself back into position, rather than a tool that cuts away the crimp. Cutting out the crimped portion of the primer cup in the case head removes some case material and, therefore, "officially" changes the dimensions from the SAAMI spec. Tools that swage or push the crimp back into position do not *remove* material, rather they *displace* the material back into its original location, as they make the case head and primer pocket ready to receive the new primer. There will be more on primers and their types a little later on.

PRIMERS

Though reloaders often begin their reloading process with their bullet selection, the primer is really where it all begins in the reloading process. There are a number of different primer sizes, and there is variety in the strength of the flash, or energy, of both different sizes and makes of primers. One thing to keep in mind as we examine this topic is the French word *brisance*. This word is used to indicate the amount of flame generated by a specific primer and it's terminology you'll sometimes encounter as you continue your reloading education.

The primer is the component contained and held in the center of the base of the metallic cartridge case. It contains an explosive compound. When struck on its metal side by the firing pin in your gun, it will ignite the powder sitting in front of it within the case, building up pressure to send the bullet out of the case, down the barrel, and out to your target.

In general, you will encounter two *styles* of primers. The first primer type is the Berdan primer, named for its designer, Hiram Berdan. Berdan primers are sometimes found in surplus ammunition. This primer type has two flash holes through the case head in the base of the primer pocket, with a pronounced bump between them. This bump replaces the self-contained anvil in the Boxer primer, which I'll get to in a minute. In a Berdan primer, the priming compound is crushed and ignited between the cup and this protruding portion of the case.

In reloading American metallic cartridges, as well as most other modern cartridges, the primer used will be the "Boxer" primer, so named for its inventor, Edwin M. Boxer. These primers are best distinguished by the single, central flash-hole in the case head's primer pocket. They also have a self-contained anvil, which allows for the crushing and ignition of the priming compound. Boxer-primed cases are the reloadable cases we will discuss throughout this book.

Standard Boxer primer sizes are Small Pistol, Large Pistol, Small Rifle, and Large Rifle. You also will find magnum versions of these for specific cartridge and load applications. Each primer type is made specific to its intended purpose, with rifle primers having heavier metal in their cups to withstand greater pressures. Magnum primers also have heavier cups, though to a different degree. *Never* substitute another primer type, size or brand for the specified one in the loading data manual, as this can create both a functional and a safety problem with the finished cartridge.

To determine the correct primer, you should consult the load data from your powder and/or bullet manufacturer's reloading book (more on that in the bullet section of this chapter). That data will list not only the size, but the brand of primer used in each and every load development and pressure testing. This is critical, as the same size primer from different manufacturers may have a different level of *brisance*. (And now we have used the cool word we learned in context. Consider yourself less of a novice!)

POWDERS

Modern smokeless powders burn at a predetermined rate based on their makeup and coatings. At their most basic, powders are generally found in two chemical forms, one based in cotton fiber, the other based in a cellulose or paper-like materials. In their simplest forms, your gunpowder is made by soaking one of these base materials in nitroglycerine and then shaping, forming, extruding, cutting it, etc., to produce the desired structure that will not only flow through the powder measure and into the cases you're reloading, but one that will also have an impact on the rate of burning (that burn rate is based on the surface area of the powder). Finally, a series of chemical coatings are employed that will also help control the rate at which the powder burns.

It's important to distinguish that today's modern smokeless powders do *not* explode! Properly and technically, they burn at a high rate. This burn creates immense volumes of gas that force the bullet forward from the case in which it's held and propel it down the barrel.

Now, an library of books on powder history, design, and manufacture could be created. There are, literally, hundreds, if not thousands, of powder "numbers" or names that have been sold for

There is an almost endless variety of smokeless gunpowders available to the reloader.

Reloading manuals are generally available from the big names in component manufacturing, like Barnes, Nosler and Sierra, but also are available from component and press makers, such as Lyman, as well as powder manufacturers. Pick just one to start and get to know your press before you start loading for everything under the sun.

reloading over the last century. For the sake of this discussion, we can keep things fairly simple.

The Internet has, in many ways, replaced the "basic" and free powder manuals of old, though most manufacturers do still produce these, as do many bullet manufacturers. I have found the Hodgdon Load Data Center at www.hodgdon.com to be a fast and easy resource for finding the correct powder for your reloading application. After entering the cartridge you wish to load into Hodgdon's search base, the system will generate a printable table of powders, primers, and specific loading data for the bullet/s you choose. The site is so functional and provides such worthwhile information that I added a long-range Wi-Fi link in my garage loading room, just to have access to this information. Hodgdon-branded, IMR, and Winchester powders are all listed on this system.

When I consider a powder, I look at the broader scope of cartridge applications. This helps to control costs and keep space considerations in check, for instance, if one powder will serve the loading needs for a number of cartridges I use.

BULLETS

Bullets for metallic cartridge reloading come in as dizzying an array of shapes and sizes as do powders, and in far more than the variety of rounds themselves, because nearly every cartridge is capable of firing more than one bullet weight and shape. So how do you choose? Where do you start?

You have to start with the basic bullet shapes. They are as follows, with abbreviations you'll likely see in reloading recipe data:

- Full Metal Jacket (FMJ)
- Round-Nose (RN)
- Jacketed Hollowpoint (JHP)
- Hollowpoint (HP)
- Wadcutter
- Semi-Wadcutter (SWC)
- Solid
- Boat-tail Hollowpoint (BTHP)
- Solid-Core
- Pointed Soft-Point (PSP)
- Jacketed Soft-Point (JSP)
- Ballistic Tip
- Semi-Point (SMP)
- Lead Round-Nose (LRN)

Sometimes you'll see some of this nomenclature used together. Thus, a boat-tail jacketed hollowpoint is expressed BTJHP. Now, again, these are the basics of bullet shapes and general design. Every bullet maker has its own propriety bullets beyond that, something that offers the reloader dozens of options for even a single caliber. For instance, in 9mm, Sierra offers six varieties, Hornady 14—that's 20 different toppers for your 9mm Browning Hi-Power, and that's just from two component makers! Add in Barnes, Winchester, Remington, and bullets from easily enough other makers to make you take off your mittens to count, and you can see how much variety you can experience with reloading.

But back to the question of how do you choose. First consider what it is you want to reload for? Are you punching holes in paper at 50 or a 100 yards, or do you need to knock over steel at 500 yards? Are you loading for self-defense or for taking down a big-game animal? Once you've narrowed the available field of bullet choices by purpose, the easiest way to choose from that group is to pick one reloading manual from one of those bullet makers (Barnes, Sierra, and Hornady manuals are terrifically composed), and study the recipes for the caliber you want to reload. In each of those books you'll see the powder and primer recommendation and other crucial information to make it all go together correctly; and, because the powder is matched to a particular load with a particular bullet weight and shape, this is where you'll make your powder and primer selections, too.

These books are *essential* to reloading, and they are something you will use, *must* use, every time you begin to reload. This is true whether you've reloaded your pet load 200 times or you're experimenting with a new load or caliber. The reason, of course, is safety. Reloading isn't like throwing together a grilled cheese sandwich, which you've made a thousand times and which doesn't require a cookbook. You're handling explosive and near-explosive components (primers and powder) that will do what they do in a container of metal (your gun), and over- or under-loading your round can produce catastrophic results—and I do not mean a bad group on your target.

If this is your very first time reloading, pick just one manual from one bullet maker. Pick one bullet and load you believe will work well in your gun for your intended shooting purpose, likely one comparable to the factory load you've been shooting, and experiment. Indeed, experimentation is a good deal of what reloading is about. Just be careful as you go. Work up one load at a time. Do not switch loads in the middle of a session. Separate your loads, label them accordingly, and keep records of how they performed in your gun. That's how you reload successfully and stay safe doing it.

SAFETY

Safety is always the first consideration, when we are reloading any cartridge. Like all activities, there is some risk involved, but this risk can certainly be minimized, even nearly eliminated, with a good routine of safety practices.

Safety glasses should be worn during any reloading activity. The work area should be clean and well organized, so that handling accidents do not occur. Please take time to consider the movements required in the reloading process and then place the components you will handle—cases, bullets, powder, and primers—in convenient and easy to reach locations. Reaching across the workspace leads to spills and accidents.

There should be no food in the area. Depending on what you are loading, there can be trace amounts of residual lead. We do not want our food to be contaminated with lead, nor do we want our components contaminated with food residues. Always wash hands immediately following your reloading session to be sure you do not carry lead or any other substance to a food source.

There should be no smoking—at all, period. Powder and primers can be accidentally ignited, and while powder burns rather than explodes, it does burn very hot and very fast. Think of powder as you would any highly energetic flammable substance. Store it in its proper container and follow the directions supplied by the manufacturer at all times.

Your handloaded ammunition can look and perform every bit as well as fresh factory ammunition—and often even better!

3 Understanding the Brass Case

Having looked at the more volatile components of powders and primers, let's take a deeper look at the thing that holds them: your brass cases. While bullets, powders, and primers merely requiring careful handling, your cases require specific preparation in order to be reloaded and, thus, this topic deserves its own chapter—yes, it's that important.

The cases you'll be using have not only been fired, but put through a tremendous series of very rapid stresses. During loading and firing, a case will have the pressures exerted on it taken from atmospheric to, in the case of the .270 Winchester, for instance, over 70,000 psi, and then back to atmospheric, all in milliseconds. The case will also have a bullet wrenched from the grasp of its neck, and firing will have stretched the case.

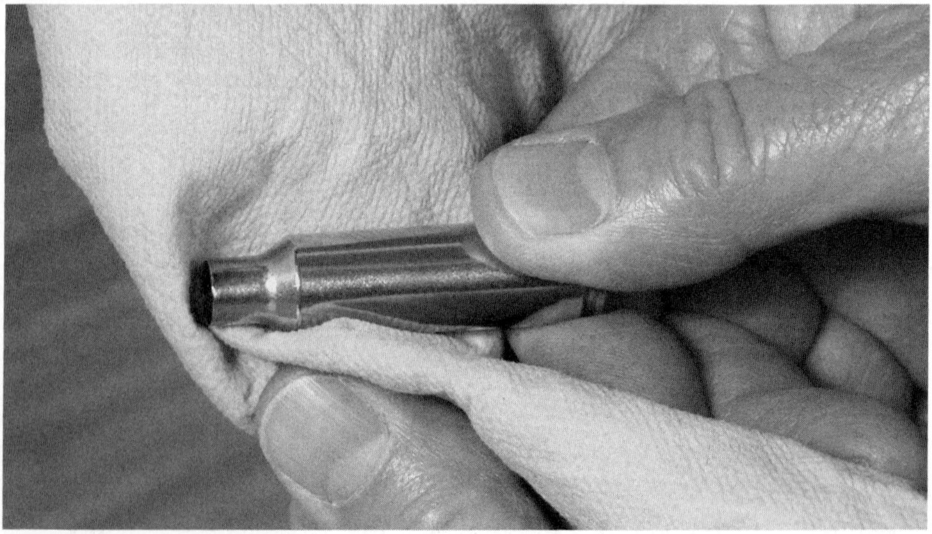

A bottleneck case for reloading. Sorting, inspecting, and cleaning your brass cases are important first steps in the reloading process.

CASE STRESSES

This is a good time to discuss just how the case lengthens in a bottleneck cartridge during the firing process, and also to outline some of the telltale signs of the stresses to look for in the inspection and sorting process of the brass. I think you'll find this interesting, as it will likely be just the opposite o)f what you would expect.

When you pull the trigger on your gun and the hammer "drops," the firing pin, striking the primer, drives the case forward in the chamber until the shoulder of the case comes up hard against the chamber's shoulder. This serves to better center the neck of the case and the bullet in the bore axis, which assures greater accuracy potential. As I previously explained, the pressure curve begins to build very rapidly before the bullet begins to release, or "pull," from the neck. This pressure is distributed within the interior of the case, causing the case to expand radially. What comes to pass here is that the case thins as it progresses from its head to its neck as a normal part of the brass "drawing" process.

The thinnest part of the case below the shoulder is really and truly immediately below the shoulder, i.e., the body junction. Here the brass expands first with the greatest force, gripping the walls of the chamber with intensity. As pressure increases, the case body from the shoulder junction back towards the primer end expands (at a linear rate) up to the "web," the thick section just above the case head. At the web, the brass begins to become too thick to expand radially; the pressure cannot overcome the case's structural strength at this location.

Remember that, at this point, the case head is no longer seated against the bolt face, as it was driven forward (of course) with the rest of the case by the action of the firing pin's impact. As the pressure of the second or main pressure curve mounts, though, the rear of the case, wanting to expand to fill the complete cavity of the chamber, now moves back until it impacts the bolt face. This causes the body of the case to stretch, increasing its length, but the stretching isn't uniform. Because the thinner sections of the case body are firmly gripping the chamber walls, the vast majority of the case stretching occurs adjacent to the web body junction where the transition from the thinner body abruptly changes to the much thicker and non-expanding web section of the case.

This is where the potential for a problem known as "case head separation" begins. Signs of this problem manifest as a clear, bright ring

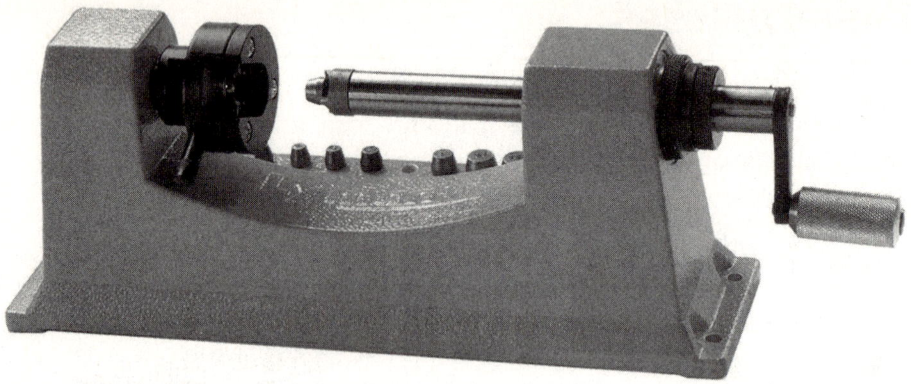

Crank-turn case trimmers, like this one from Lyman, can make quick work of this necessary case preparation chore.

around the case, the ring visible just above the case head. Case head separation is a very dangerous situation, and its signs should be looked for in your case inspection process prior to beginning the reloading process.

The second area in which we see the case lengthen is the neck. This is due to the brass flowing forward from the pressure on the interior of the shoulder and the bullet pulling out of the neck with great force. Brass flows in a plastic-like way when exposed to great pressure, and this flow is amplified by the resistance to the force of pressure offered by the constriction of the shoulder. We are, literally, *dragging* great pressure along the surface of the interior of the cartridge case, causing it to be pulled along like the wind pulling and moving the surface of water. This is manifested by an increase in the length of the neck of the cartridge case.

Just like case head separation, case neck lengthening can create a dangerous situation, if the neck exceeds its SAAMI specification. The danger here results from the larger diameter of the neck moving to a point forward of its allocated dimension in the chamber. Ahead of where the neck is designed to be, the chamber steps down for the lead of the rifling. This portion is what the bullet diameter should be *without the addition/interference of the thickness of the case walls*. If the case neck is too long, you then have the larger diameter of the neck and the bullet in this smaller area. If you have this problem, when closing the bolt, you effectively create a hard metal crimp. That crimp pinches and holds the bullet and will not allow it to properly pull or move from the case when the cartridge is fired. In the worst case, this can cause a

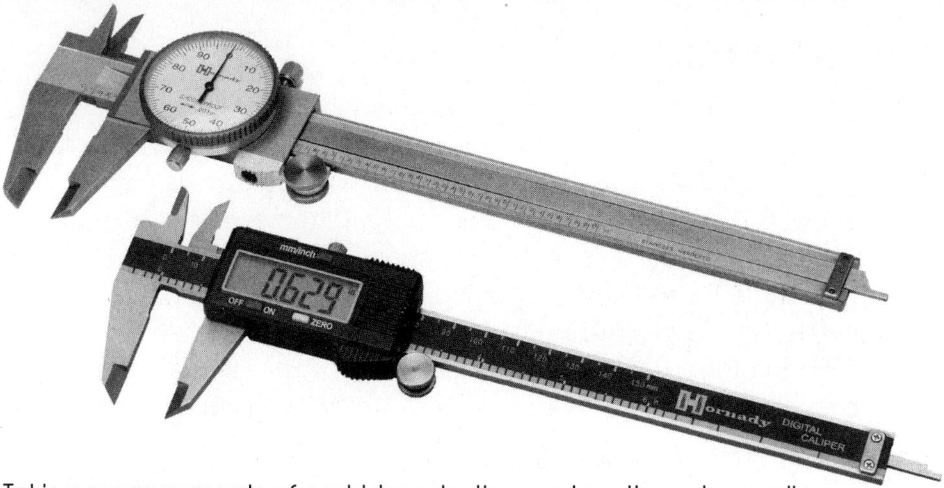

Taking measurements of cartridges, both case length and overall cartridge length, is an important part of successful reloading. Calipers are the way to do this, and they come in both dial and digital versions.

catastrophic failure, where the gun literally ruptures due to exceedingly high pressures resulting from a barrel, which is, effectively, plugged. This great pressure can reduce the receiver and chamber area to a mass of shards and splinters.

Now, we generally use brass as the material for cartridge cases, because of its ability to "obturate." Earlier, I used the word "plastic," in describing how brass flows as it stretches. That description wasn't, obviously, about the chemical makeup of the brass, but rather about how brass has the ability to change, including a return to an original shape or size. "Obturation" is the technical term for the way brass can expand, keep its overall shape and, then, after pressure is released, spring back to dimensions smaller than when it was fully expanded, all while continuing to maintain its general original shape. The ability to obturate is necessary for the case to be able to expand to grip the chamber walls, seal the chamber from the unintended escape of hot gasses, then spring back slightly, where it reduces in size so that it can be easily extracted.

Why this discussion on obturation? Because this process has an impact on how we size the case. As we constrict and compress the case, it will want to grow slightly larger when the compressive stress is removed. In other words, obturation works on brass cases both when the cartridge is fired and when the case is being worked over in the reloading processes.

BRASS SORTING

Now that you know about case lengthening, brass flow, and all the other things that happen to a metallic cartridge case when it is fired, the first good practice in your reloading process is to always inspect and sort your brass. First, you should segregate those cases that came directly from your firearm. Second, you should sort your cases by brand of manufacture. This process will reduce the number of future adjustments, as you progress through the reloading steps.

I am fond of using inexpensive, GladWare-type plastic containers in which to sort and store my brass. They are available in a number of sizes and also offer a good option for long-term storage, due to their airtight seals; moisture and other contaminants stay on the outside. I once opened, with great excitement, a box of brass kindly given to me, until I found it had been a condo for mice at one point. The box was filled with rather nasty stuff mixed among the sorely corroded remains of the cases. Completely unusable (and quite smelly), it went straight to the trash. But back to the sorting of your brass.

As I said, you should sort your brass by manufacturer, after you have separated the brass that was fired from your gun from those that were otherwise procured; remember cases fired in firearms that are not the gun you are reloading for require some additional prep and care. The differences in chamber specs between guns produces fired brass that is less than well suited to *your* firearm. As you sort your brass, you should be looking for obvious signs of trouble, such as cracks or splits. All cases displaying such wear should be immediately discarded. Especially check the case necks carefully, as these are areas where cracking can begin. Next, inspect all bottleneck cases for signs of case head separation, looking for that bright ring on the brass 1/8-inch to a ¼-inch up from the case head. If you find such cases, again, the answer is to simply discard. Personal safety is not worth the few pennies a used piece of brass may cost.

BRASS CLEANING

With fired (and, so, obturated) brass, the first step after inspection is cleaning. Primer ignition and the burning of powder leave behind residues and carbon. A successful reloading process must deal with the removal of such residue and debris, so as to provide a solid foundation for the resulting cartridge to function properly and safely.

Cleaning can be as simple as a cloth to remove excess dirt and residue from the outside of the case and a brush to clean the inside of the neck. This latter is important, as it's this portion of the case that holds the bullet under tension and releases it cleanly and (hopefully) equally around its circumference. Using a neck cleaning brush is the simplest and most basic method of cleaning the case.

If you want to clean the entire case, there are a large number of tools available from many manufacturers that will perform this job nicely. They range from simple plug-in tumblers that use a rough polishing media on up to the more expensive ultrasonic systems. Ultrasonic cleaners have a very good reputation among some very proficient volume reloaders I know, but these units use expensive chemical solutions and, in my humble opinion, can be a bit messy.

I may be showing my age, but a good quality vibratory tumbler and good old corncob media has served me well for more than 20 years. A basic kit is available from many manufacturers. Mine wears a Midway USA label and has been rock-solid through many thousands of cases.

A quick side note about media storage, should you choose to go the tumbler route (which isn't a bad idea at all). I buy my tumbling media in bulk and keep the extra in washed and dried one-gallon milk containers with screw-on lids. My reloading shop is in the garage on my farm, and mice just love corncob media, especially in the winter. The screw top and good plastic seal out the little buggers and also prevent spillage.

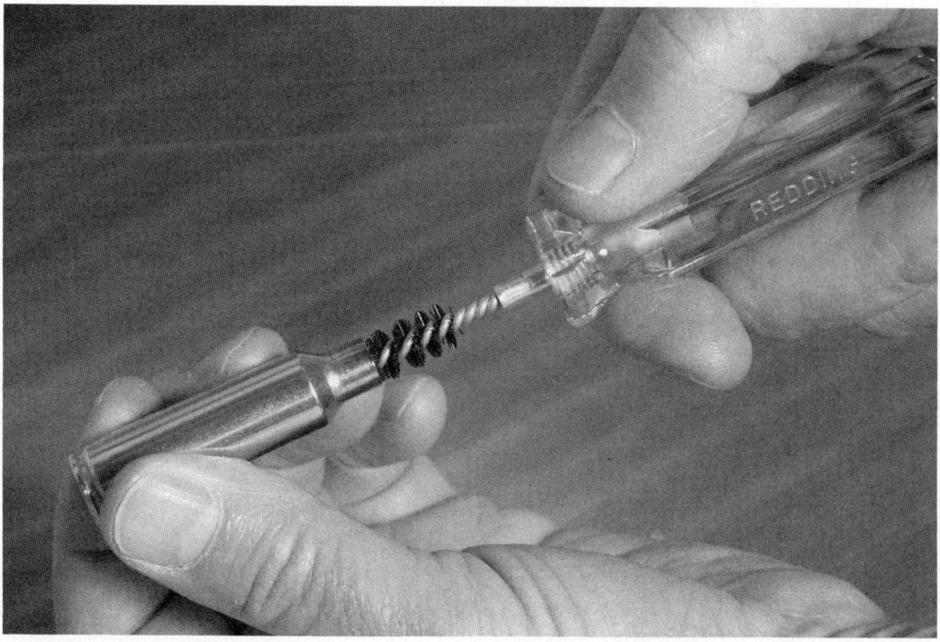

Cleaning the inside of the case neck with a neck brush.

An electric, vibratory tumbler and media will be necessary items for the metallic cartridge reloader. They come in a wide variety of sizes and hopper capaci ties. The one below is from Hornady. Tumbling media, the stuff that cleans and polishes the brass in the tumbler, also comes in many forms. The author prefers corncob tumbling media over others, but notes that you must keep it sealed up—mice love this stuff!

4 Reloading Presses for Metallic Cartridges

Metallic reloading presses come in a number of forms today. In general, there are three distinct categories of presses: the single stage, the turret press, and the progressive press. Within these major categories there are a wide range of derivations with an equally wide range of prices. For the purposes of this book, I'll focus on only those presses that take the standard 7/8-inch/14 TPI (threads per inch) as made by American manufacturers, as well as standard shellholders, on all but progressives. These presses will work as described within the pages of this book without the need for text dedicated to the differences in processes encountered in specific press models.

SINGLE-STAGE PRESSES

Single-stage presses are the most basic. In their simplest form, they consist of a movable ram, which accepts a shellholder, and a threaded 7/8-inch/14 TPI hole centered directly above the ram for the insertion and retention of a die. These two features, along with a mechanism to move the ram upward, are all that are needed to make a reloading press for metallic cartridges.

A single-stage press can perform one operation of the overall reloading process at a time, on one cartridge case at a time, thus the moniker "single-stage." Variations in these presses are many, with differences in materials, compound leverage size, weight, and design. In the end, each type serve the same general functions, but the differences between makes will allow some single-stage presses to have a greater range of capabilities, due to strength of those press designs coupled with the strength of the material from which they are constructed.

That may sound vague, but these factors come into play when you are full-length sizing large cases that have a great deal of surface area, or cases that have been fired in a chamber with headspace on the longer

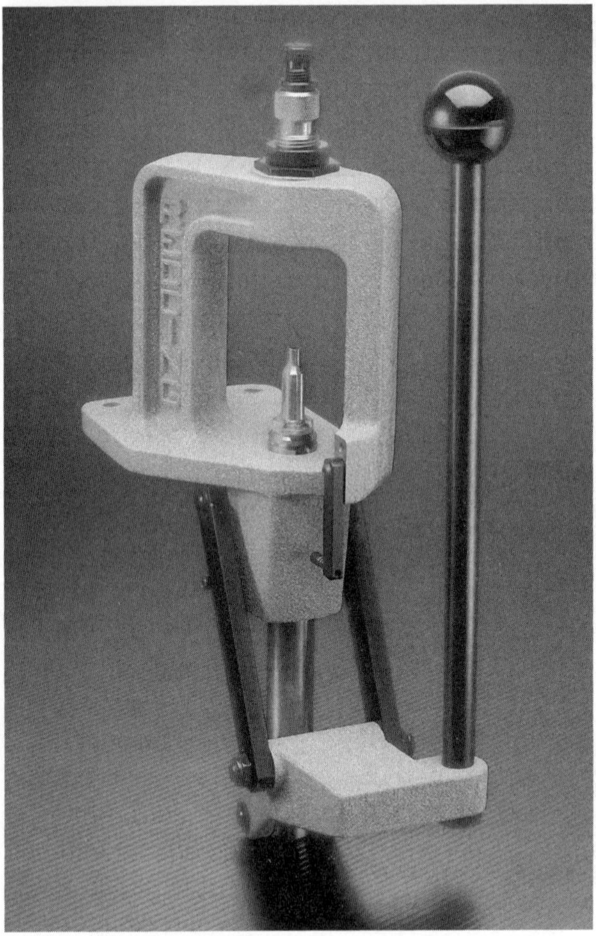

Modern single-stage presses are strong and accurate, producing quality loads for both rifle and handgun cartridges.

side of acceptable. In these instances, power and strength are needed to push the shoulder back to a position to be used in any SAAMI specification chamber of the caliber. To do this properly, and especially in combination, requires a very strong and rigid press with great mechanical advantage.

The most important feature of any press is its ability to maintain its dimensional stability and not deflect in any critical dimension of alignment under the maximum load it can generate. That sounds like a mouthful, but such a deflection would create some form of misalignment and, ultimately, result in a loaded round having some measure of deficiencies.

There are two philosophies to creating single-stage presses in this regard. The first method is to balance the force available to the size, capability, and limitations of the press frame. The second method is to build an incredibly strong press through the use of large iron castings, and couple that strength with engineering that maximizes the reduction of any form of dimensional shifting under extreme load. This is not to say that all presses should be massive and have immense power. Some presses, for instance, slant to the lightweight, for use as a range press at a benchrest match, where neck sizing only and bullet seating are required. Another example would be a press that's intended to be used where there was limited space available. At the other end of the spectrum, there are massive presses designed for the large African magnum rounds and big, straight-wall cases like the .45-120 Sharps, where tremendous forces need to be generated. Single-stage presses to load these giant cartridges are also a favorite with those who form their own cases from other calibers of brass or basic brass available from brass makers like Jamison International. (Case forming is an advanced reloading practice and, so, beyond the scope of this book, but it's good to keep in the back of your head for future reference the knowledge that some obsolete cartridges and many wildcat rounds must be made by forming, as new cases or loaded ammo have not been available for many years.)

THE C-FRAME SINGLE-STAGE PRESS

The simplest single-stage press is known as a "C-frame." These are named in this manner, as the backbone of the press is "C" in shape, with an open front. The ram is at the bottom of the press and the die station is directly above. These presses are very basic, but offer easy entry for your cases to the shellholder, due to the open front design.

C-frame single-stage press rely on the rigidity of the material and, in some cases, the structural design of the press' casting to produce the best possible dimensional stability under the heavy load needed for a full-length sizing (which you can see from the previous chapter is often a required step in your overall reloading process), but they're available in a wide range of sizes. Lee Precision, for instance, builds a simple and inexpensive bench-mounted C-frame press called the Lee Reloader Press. That same company also builds a unique hand-powered C-frame press called the Breech Lock Hand Press. Both have compound linkage and are capable of good full-length resizing on smaller cases. These presses are inexpensive and ideal for a new reloader who has limited space and a limited budget for their initial reloading setup.

At the extreme other end of the C-frame spectrum, Redding builds its UltraMag

A unique hand-powered press is the Breech Lock Hand Press offered by Lee Precision. Note the C-frame design.

The Redding UltraMag Press is a modified C-frame design. Note the heavy backbone (the massive amount of metal attached to the bench) and the links (the long, flat, black bars) that extend to the head. The design removes the main shortfall of some simpler C-frame presses, which is flex under load, i.e., movement of the press and misalignment of components and machinery during press operation.

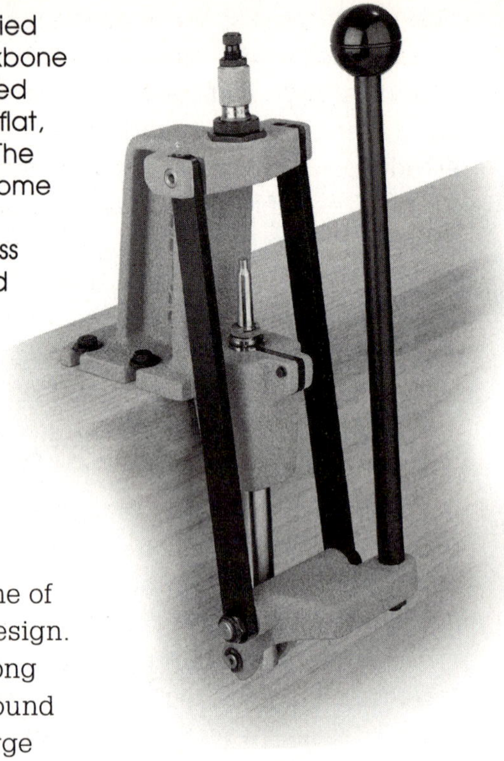

Press. This is a modified C-frame using extremely long side link bars. These bars actually attach to the "head" of the press, balancing the load while still allowing for the easy access to the case, which is one of the biggest pluses of the C-frame design. Redding's press has an extremely long ram stroke and high levels of compound leverage, making it ideal for very large cases. Additionally, while it ships with an installed $^{7}/_{8}$-inch/14 TPI bushing, it will accept 1¼-inch/12 TPI dies with the bushing removed. Redding also offers a 1-inch/12 TPI bushing to support large-caliber dies of that dimension. This press will easily handle cartridges as large as a .408 CheyTac.

THE O-FRAME SINGLE-STAGE PRESS

As press design evolved, the next natural step was to design a press with greater strength and rigidity. While traditional C-frame presses were fine for many smaller cases, they could deflect under the heaviest loads. The advent of the O-frame press was a benefit to reloaders of larger cases, which require heavier sizing pressures.

The design of the O-frame press is not really a round "O," but rather one that more closely resembles a rounded-corner square, with the ram located centrally in the bottom of the "O" and the die attachment point directly above on center. This design improves the elimination of distortion under a heavy load and keeps the die and ram in square, true alignment.

At the same time the O-frame gained popularity, the lower portion of presses began to evolve, with heavier rams and changes to linkages providing greater mechanical advantages. Early linkages were generally a short, toggle-style link contained in a milled slot within the body of the ram at its base. There were limiting factors on power with this arrangement, due to the restrictive lengths and location of the linkage and its attachment points. Progressive linkages improved their mechanisms, with heavier links mounted on the outside of the ram and with hardened, transverse cross-pins bisecting the bottom of the ram. An independent lower link, usually of cast iron, allowed for a progressive compound linkage to develop much higher forces at the apex of the ram's travel. This also allowed for a longer area of traverse to be established in the lower portion of the press frame through which the ram passed, and it added strength and increased bearing surface, eventually ushering in the era of ever increasing compound leverage.

This is the common form of lower link arrangement seen today. The design generates high levels of force through its travel, culminating at "cam over" at the apex of the ram's travel. With these improvements, ram diameters also grew, providing both greater strength for sizing and a further increase in bearing surface as the ram passes through the press frame. This single improvement was a major factor in reducing ram run-out, as the longer and greater diameter of the bearing surface limited the amount of play in the ram during its traverse.

Redding's Boss press is a great starter press and available in a Pro Pak kit for the beginner.

O-frame presses are made in a variety of sizes and from a variety of materials. If you're leaning towards this press design, you should consider the available size of your mounting location, as well as the intended uses and the weight of the press. Lee makes aluminum-framed O-frame presses in multiple sizes, as well as a large, cast iron O-frame. Redding makes its Boss, Big Boss, and the Big Boss II press, which are also cast iron and extremely strong. The universe of O-frame presses is large, and your options will find expansion through diverse markets like E-Bay, where many "classic" presses can be had at good starter prices. When buying used, just be sure the press will do what you need and that the ram does not have lateral "play," which will result in poor alignment.

A turret press holds multiple dies that index into position for use, but these are still considered to be single-stage presses.

TURRET PRESSES

Turret presses occupy a unique position in press design. They are, at their heart, single-stage presses, as they perform only one operation of the reloading process with each stroke of the ram. But, by using an indexable turret head, various dies can be mounted at one time, thereby covering all the needs of one caliber and, in some cases, even multiple calibers. A turret press is generally more expensive than C-frame and O-frame presses, but their inherent flexibility and the fact that they are truly single-stage in operation still make them a sound choice for new reloaders.

The design of the early turret press likely traces its roots to the C-frame press. It has a base and backbone casting, but they lack the top curl of the "C." Instead, it has a turret head mounted to the top of the backbone, that head accommodating a number of dies. Each die mounting location will precisely align on center with the ram, as the turret rotates. Lyman was a big mover in the turret press business and made a number of models through the years, including the Spar-T and the All American. Later, Lyman produced another turret press, the T-Mag. The T-Mag was my first "good" press and served me well. That model has since been updated and improved even further to become the T-Mag II.

Redding Reloading is another company with a history in the turret press market, beginning with its Model 25, a six-station setup with a simple linkage that, despite its lack of complexity, was still a great press. Today, Redding makes the massive T-7, which will hold seven dies and weighs more 27 pounds. The big T-7 features compound leverage, a through-ram primer disposal system, and the ability to interchange turrets with the dies remaining in proper arrangement.

The Redding T-7 is the press I use today, and I love it. The seven die stations are a great advantage, as I can setup a three-die rifle set with sizing die, seating die, and a taper crimp die, plus a four-die setup for

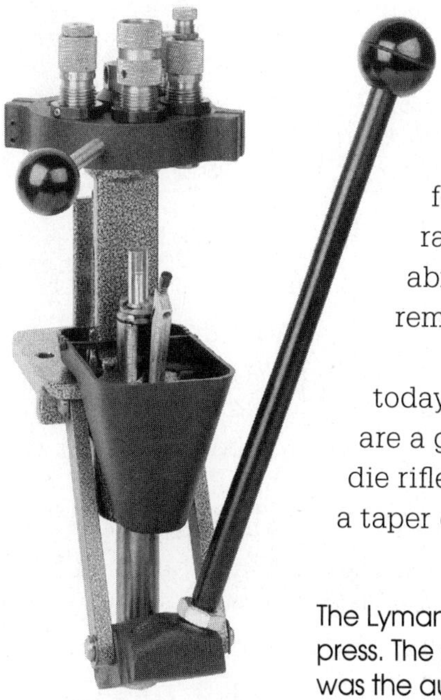

The Lyman T-Mag II is a solid-quality, cast iron turret press. The press' immediate predecessor, the T-Mag, was the author's first "good" press and served him well.

handgun cartridge reloading that includes sizing, expander, seating, and taper crimp dies. So arranged, all my needs for .223 Rem. and .45 ACP are there and I have no need to change dies or their arrangement. Also, I can also switch out a turret and be ready with other calibers very quickly. This illustrates the true beauty of a turret press and its ability to retain a number of dies in proper position, thus eliminating the need for constant die change and setup routines that are inherent with the simpler single-stage designs.

The Redding T-7 holds seven dies at once and features an interchangeable turret head that reduces die setup effort and time.

There were two detrimental factors to the early turret presses that buyers of used presses should be aware of. The first is that some presses had turret heads that were not always in true alignment to the ram, and, so, they were prone to tip out of square under load. The second problem is that early turrets also lacked a detent ball system for precise alignment of the die location to the ram and, so, with those, it was dependent on the user to accurately rotate and align the die centrally over the ram so that the cartridge case could properly enter the die. This became an issue if working with large-caliber cases; if a die was misaligned, it could crush the case during die wall contact with the case mouth.

Another feature to be aware of is that some early turret presses had small-diameter turret heads. This may give you the impression that less leverage would be needed, which would limit tipping, but, in actuality, the smaller designs did just the opposite. A larger diameter allows for a support strut to be added to the frame casting. The support

A turret support coupled with CNC machining and a heavy backbone on the Redfield T-7 turret press virtually eliminates turret tip under a heavy load.

strut is located opposite the ram and directly behind the central supporting shaft of the turret to balance any deflection occurring under load. The greater the distance between the central shaft and the support strut, the less the possible deflection with the same amount of clearance to allow rotation of the turret head. The central shaft, in essence, becomes a fulcrum and the limit of travel at the end of the ram stroke, hence, the tipping moment is constrained by the impingement of the rear of the turret head upon the top of the support strut. This design also allows for a placement for a detent ball at some distance from the central shaft, thus creating a more precise alignment of the die location above the ram's center.

Redding designs its turret presses to actually tip very slightly, but into square rather and out of square. It is an interesting design and one very possible to achieve with today's CNC machining technologies. The head is of a large diameter to hold the seven die positions, and a very stout spring powers a large-diameter detent ball, which recesses into position on the bottom of the turret head. Redding also places the detent ball in the center of the support strut as a clean design feature.

I will admit I am a big fan of turrets and have owned four different models, two from Lyman and two from Redding. All have given me very good service and good handloads, as my needs progressed.

PROGRESSIVE PRESSES

Progressive reloading presses are very different from what I have discussed above, as they uniquely do a number of functions with each stroke of the ram. I have owned progressives from Lee and I still own a Star Universal setup for .38 Special bull's-eye loads. The progressive press fills a need for high volume, good-quality reloading, in that it performs the many individual stages of the reloading process all at one time, via different "stations" on a rotating base. While one station primes, another sifts in the powder charge, and yet another seats and crimps the bullet, all with one pull and lift of the lever. In general, the progressive presses of today will produce ammunition as good as any American ammunition company, if you are a diligent and competent reloader.

I've saved the progressive press for last for a couple reasons. One, the practice of reloading couldn't have evolved to include the creation of progressive presses without the individual stages inherent in single-stage and turret presses. My second reason, and more importantly, is that, in the case of the beginner, for whom this book is primarily aimed, I am not one to recommend a progressive as a starting point. As a new reloader, you need to focus on perfecting each action in the process, something certainly better enabled with a single-stage or turret press. Until you're familiar with some of the ins and outs of the reloading process, the many operations taking place all at one time on a progressive press can cause both frustration and a good deal of wasted material. Like all learning, quality is the first priority. Quantity and speed can wait.

As a final note on this topic, your single-stage or quality turret press will always serve you, even when you have the progressive cranking out high volumes. Why? Because your precision loads for long distance will always be loaded one at a time. I am amazed when I read the exchanges on competition forums, where serious shooters who own sometimes multiple progressive presses relate the tales of their most critical ammunition for their most important events being loaded on a turret or a single-stage. That says a lot about the value of owning and using a single-stage press, even when you've moved on to a progressive for your high-volume shooting needs.

Take the time to learn, understand every step and process fully. Build good reloads that function and are accurate, then graduate into all that the ever-evolving art of true handloading and customization can offer you and your shooting experiences. Grow in the process and find your perfect loads. Get the perfect bullet seating depth and make your rifle sing with groups that continue to shrink, as your knowledge and experience continue to expand in the art of handloading.

The following chapters in the next two sections explain the individual steps to loading bottleneck and straight-wall cases. In these chapters you'll see that the steps to loading are explained as they would be in a single-stage press setup. Part of the reason for this is because the steps need to be done in a specific order—for instance, it should be obvious that you can't charge your case with powder without a fresh primer first being seated—A second reason for addressing these steps as done in a single-stage press is so you have a clear idea of the care necessary in taking those steps, and that's something most easily seen and explained with the use of a single-stage press setup. Again, if you're a first-time reloader, the single-stage is where I recommend you begin your new hobby.

PRESS ROUNDUP

Most of today's reloading press makers have been at their trades a long time. In fact, you don't often see new press designs come on the market, though every year there seems to be a fresh bevy of accessories that run the gamut from basic to highly specialized. It would take another volume to cover all the reloading equipment that's out there, and while there's a reference list of not only press but ancillary equipment and component suppliers at the end of this book, I think a brief listing of the presses you can expect to find online and in retail stores is useful.

DILLON: This is one of the newer players on the scene, though, with more than a quarter century behind its history, it's hardly a baby anymore. Dillon offers three centerfire metallic cartridge reloaders, all progressive presses. The Square B model is the one many new loaders start out with, as it is the smallest and most compactly arranged of the three. The Square B is also intended strictly for handgun cartridges, and the press comes with its own specially sized dies, making it ideal for the avid hobby shooter who has only one or two handgun calibers to deal with. The RL550B is the middle sibling, offering the capability of loading more than 160 rifle and handgun cartridges, plus there's a magnum powder conversion available for loading such behemoth cartridges as those in the Weatherby and Nitro Express families. Though a progressive, this press indexes manually, so, if you want a press that does it all, you're going to go straight to the XL650. This

A progressive press, such as this setup from Dillon, can really satisfy the needs of high-volume shooters.

big boy auto-indexes, boasts a cycle rate of 1,000 rounds an hour (for those who know what they're doing and have plenty of hours in front of the reloading bench under their belts), and has a number of fancy add-ons available, such as a powder level sensor.

HORNADY: Though perhaps best known for its ammunition, Hornady also makes three fine reloading presses. The Lock-N-Load Classic is a sturdy single-stage press, simply designed, but with nice features like automatic primer feeding and the capacity to switch dies between calibers in just about the blink of an eye. Hornady also sells the Classic in kit form, complete with powder measure, a handheld priming tool, three die bushings, powder trickler, case lube, and a host of other essentials. When you're ready for a progressive, you'll look to the company's progressive Lock-N-Load AP. Recently upgraded with a new cartridge eject design, this auto-indexing press also gets a case activated powder drop and quick-change powder metering inserts as standard equipment. Hornady's final press is its .50 Cal BMG press, a single-stage specifically designed to handle the massive .50 BMG—Browning Machine Gun—rifle cartridge.

Lee's single-stage O-frame presses come just as they are as well as in kits like this 50th Anniversary Challenger Kit that includes other necessaries and conveniences such as a scale, case trimmer, and powder measure.

LEE PRECISION: This company's been around since 1958, and you don't get to stick around that long without making quality tools. Lee offers an expansive array of presses, from single-stage models with quick-change die setups to turret presses with powder drop attachments that help speed up the process. Lee also makes two large progressive press options. Its top-of-the-line model is the Lee Load Master, a massive press that includes an automatic case feeder, allowing advanced reloaders to get large-quantity reloading jobs done quickly. It's particularly helpful that Lee offers five different single-stage presses, something for nearly every new handloader or the handloader who wants a press separate from their progressive for loading smaller quantities of another round without the hassle of switching out dies and equipment on the bigger press. Lee also offers two single-stage press kits, some of the best values on the market today for the new reloader.

LYMAN: Talk about some history! Lyman's legacy extends back to the 1800s, when the company's founder, William Lyman, patented a new rifle sight. It wasn't until 1925 that the company expanded into reloading, when it purchased the old Ideal company. Fast forward nearly 100 years, and today Lyman continues to set the mark for quality on a budget with its several presses and extensive line of reloading tools and

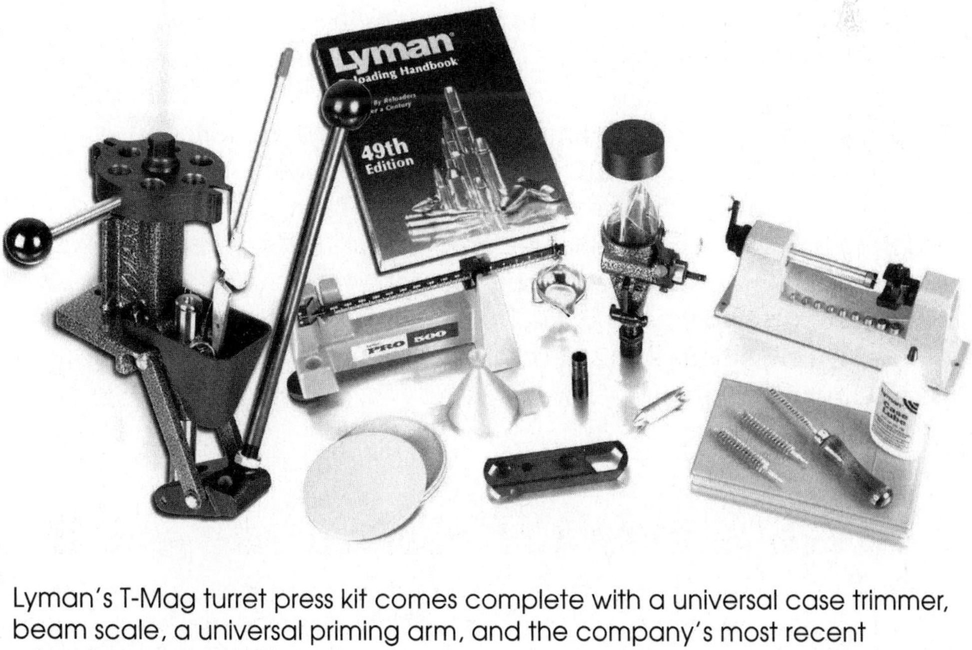

Lyman's T-Mag turret press kit comes complete with a universal case trimmer, beam scale, a universal priming arm, and the company's most recent reloading manual.

accessories. Lyman's single-stage presses, including two turret options, are most often selected in kit form, those kits ranging from basic to all-inclusive, and all include scales, either beam or digital, depending on the kit. The four presses Lyman makes are the very basic Crusher, the slightly more advanced Crusher II, and then there are the two turret presses, the T-Mag and T-Mag II.

RCBS: If you've heard the term "Rock Chucker," then you're familiar with the RCBS company. That single-stage press is one many new to the hobby of reloading begin with and one they continue to use once they move on to a progressive press. Now known as the Rock Chucker Supreme, this easy to use press isn't the only one in RCBS's lineup. In fact, the company makes three single-stage presses, including one sold completely set up for the big .50 BMG round, plus a turret press, the Pro 2000 progressive press in both manual and auto-index versions, and multiple kits for the single-stage and turret presses that include scales, chamfer/deburring tools, the Speer reloading manual, and other get-started tools.

REDDING RELOADING EQUIPMENT: In business since 1946, Redding has a stellar reputation for the quality and craftsmanship of its reloading dies, as well as its specialty tools for the advanced handloader. At the company's core, though, are its presses. The T-7 reloading press is the industry leader in the single-stage turret genre, and kits are available that include the caliber die and shellholder of your choice. The T-7 also borrows the ultra-reliable primer feeding and spent primer collection design from Redding's hallmark press, the massive Ultramag single-stage. Designed to reduce press deflection during lever operation, the Ultramag is the choice for those working with difficult and oversized cases. Too much press for you? Then the Boss, Big Boss, Big Boss II, and the kits available in this single-stage spectrum serve the needs of 9mm, .30-06, and other mainstream cartridges beautifully, and Redding's two Versa Pak kits, which have everything but the press and dies, are a simple and smart buy once you decide on which Redding press is right for you.

PART II

SECTION B: Reloading the Bottleneck Case

5 Bottleneck Case Resizing and Trimming

Your reloading press and processes require a set of dies. These reloading dies perform various operations, such as decapping (removing the old spent primer), resizing of the case, bullet seating and crimp, and more. The instruction manual that comes with your press will tell you how to install and adjust your dies for proper operation, but there are some things you should know about the dies themselves that will help those adjustments make more sense.

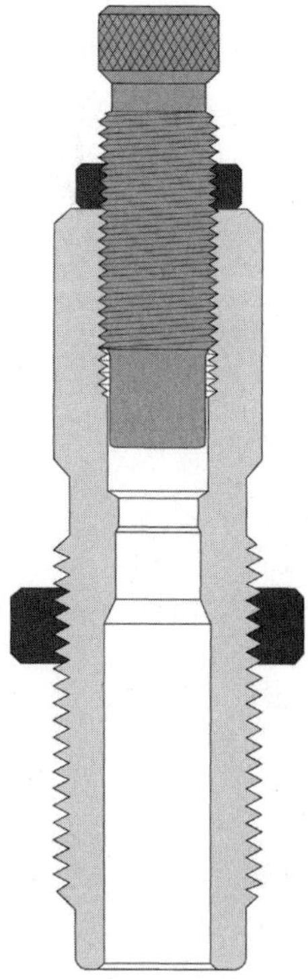
Cutaway of a bullet seating die.

Recall the conversation we had earlier about brass obturation and that the reason we use brass as the material for our cartridge cases is to allow those cases to expand and grip the chamber walls during firing, then spring back to a smaller dimension for easy empty case extraction. As beneficial as obturation is during the firing of a cartridge, it presents a problem in case resizing, which is the first step in your reloading process. You see, a resizing die compresses the metal, *reforming it to a proper dimension to fit a minimum specification chamber*. Remember, with obturation, the case being resized, under a compressive load in the resizing die, wants to

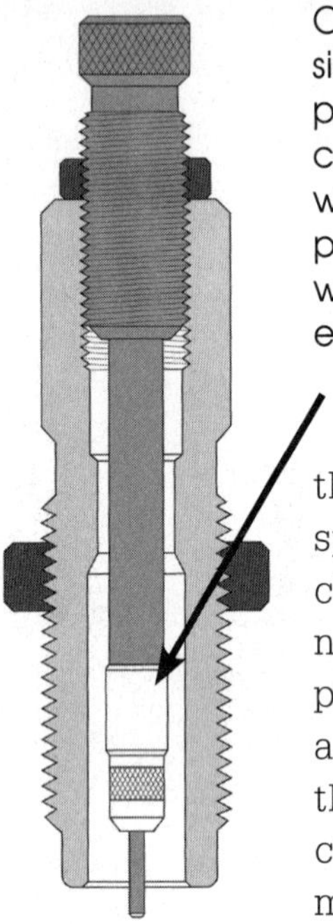

Cutaway of a full-length sizing or neck sizing die. The protruding pin at bottom center is the decapping pin, which removes the spent primer in the case. The area with the arrow is the neck expander portion of the die.

spring back to a slightly larger set of dimensions when it comes out of the die. Often, that spring-back dimension is greater than we desire.

Ultimately, this caused die designers to develop die designs that *overly* resized a case, so that the finished case sprung back to the dimensions needed for proper chambering and firing. Oh, and one note about those needed dimensions and the finished cartridge off your press: You wish to always focus on the dimensions of a minimum *chamber* specification, one that has only the slightest clearance needed to easily and safely chamber the round. We have no desire to work to a minimum *cartridge* specification, as this can and will reduce brass life.

THE SIZING DIE AND RESIZING OPERATIONS FOR BOTTLENECK CASES

Let's first take a close look at the resizing die, the first die you'll use in your press. You'll see that it has a "decapping pin" (which removes the old, spent primer from the case), and a rod to hold it, along with an "expander button," which expands the neck of a case to accept a new bullet.

As you now know, the firing of a cartridge expands the brass, so the first real reloading chore after sorting and cleaning your cases is to return them to a re-usable state. The sizing die is caliber-specific and needs to be fitted into an appropriate reloading press. We will begin the basics with a "full-length sizing die," one that resizes the neck and body, as well as pushes the shoulder back into its proper position.

A typical shellholder for mating the specific case to the press' ram (top). Extended shellholders are sometimes needed for very short cases.

Inserting the shellholder into the press' ram. This is the first step in setting up the full-length sizing die for a bottleneck case.

To do all this, the die must be properly set up in the press with the appropriate "shellholder." The shellholder is also specific, but not to caliber, rather to the dimensions of the case head. In this manner of design, families of cases can use the same shellholder. As an example, a family of cases may be one like that of the .30-06 case, whose shellholder will also work for the .25-06 Rem., .270 Win., .280 Rem., and the .35 Whelen, as well other wildcat cartridges using the 30-06 case head dimensions.

With a full-length sizing die in the proper caliber and the correct shellholder in the proper caliber family, you now have a means for setting up the die. In this we will focus on a basic setup that will allow the greatest possible sizing, assuring a good bullet tension and a smooth operation of the finished round in any chamber that falls within SAAMI specs.

Full-length sizing die setup begins by inserting the shellholder into the top of the press' ram. It is usually held by some type of spring clip that acts to both align the shellholder and retain it in its proper position.

Now you must raise the ram to its topmost position. This done, remove the major die body lock ring upward and out of the way, so that the die can be screwed into the press. Next, screw the die down until it makes complete contact with the top of the shellholder.

At this point, you must back off the ram slightly and screw in the die an additional 1/8- to ¼-turn. This creates the situation known as "cam-over" on traditional reloading presses. Cam-over is needed to ensure that the case's shoulder is properly positioned. Cam-over also correctly sets the datum line headspace for chambering, that headspace being a position just short enough to fit a SAAMI-minimum chamber. (Note that cam-over does *not* reduce the case to the minimum cartridge dimensions.) With this adjustment complete, lower the ram to its original position.

A full-length sizing die set up in a single-stage O-frame press.

Okay, you now have your sizing die installed in the press, but there's one more step before the die setup is complete. Now you must adjust the depth of the die's decapping pin travel and, therefore, the length of its rod. This is critical, so that the neck expander (part of the sizing die) does not bottom out and bend the rod. The goal with decapping is to use only enough pin protrusion to completely pop out the primer from the cup in the case head, and not to overdo this by crashing the expander into the web of the case, which will cause

Once you have your sizing die installed in your press, you'll need to adjust how for its decapping protrudes from it. You want the pin's rod to travel only enough to push out the spent primer.

damage. Depending on the brand of die set, this is done by loosening the die's lock nut and rotating the rod out or loosening the collet and withdrawing the rod upward, so that the decapping pin protrudes only slightly below the bottom of the die.

CASE LUBRICATION AND SIZING OPERATIONS

You are almost ready to size the first case, but with a steel die and a bottlenecked cartridge case, you must lubricate the brass before sizing. This step is critical to preventing the case from sticking in the die and being ruined. There are many types of lube and lube systems in the marketplace, and everyone has their favorite.

Why case lube? As the full-length die is aptly named for what it sizes and is, therefore, in contact with the full length of the case as it does its job. This involves almost complete surface contact by the die with all of the case's exterior, that die exerting literally thousands of pound of pressure that compress and push the brass into the desired dimensions and shape required for chambering; the is the "load pressure" I talked about when discussing the various presses. Here, again, we can use the word "obturation." As we compress and reshape the brass, it wants to re-expand slightly (obturate) after the force of the die is released,in the same manner that, after being fired in the gun, it contracts after the chamber pressure drops with the expenditure of the powder. During the full-length or neck-sizing operation, this obturation presses the entire length of the case against the interior walls of the die, where a lack of proper lubrication will prevent the case from being withdrawn from the die as you push the press' handle in the direction to lower the ram back to its resting position. The compound leverage of the ram has great strength in both directions, and a case can be stuck severely enough that its brass rim can be torn off in the process.

Case lubes come in liquids, sprays, and pastes. Experiment to see which you prefer and works the best in your reloading setup.

My favorite case lube, Imperial Sizing Die Wax, is very easy to use and control. Imperial comes in a tin like shoe polish. I simply open the top and rub the flat of my index finger tip on the surface of the wax until a small buildup occurs. Then I rub my index finger and thumb tip pads together, gaining an even coating on the surface of each finger. Next, I roll the case between my finger and thumb from case head to the junction of the body and shoulder. I then like to dip the neck in Imperial Application Media with dry neck lube. This coats the inside and outside of the neck with an even amount of lubrication for sizing the outside of the neck, and also the inside, for when the expander ball comes back through on the down stroke of the ram and uniforms the neck tension of the case.

Cases stick at their bottom 20-percent, as this is where the brass is the thickest. This is why I like the Imperial product, because you almost instinctively hold the case by the neck with the hand that doesn't have the lube on it. Therefore, the process of getting the greater coating of lube where it is needed most becomes automatic.

Many people like lube pad products, and I'll admit that I've had good experience with some of them. With these, you'll need to remember to keep the lube off the case's shoulder, to keep from getting dents in the case under the pressure of resizing. Another choice for lubing is spray lube, but I am personally adverse to them. While they look easy to use, I immediately find a number of issues. Most require a waiting period of 30 minutes before the cases can be used in the press, and I am not a patient person. What happens if you don't wait? Stuck cases. Additionally, as we tend to spray from the top, usually with the cases ranked, the lower portions of the cases don't always get the necessary amount of lubricant in this critical area.

The author uses Imperial's Dry Neck Lube in addition to its Sizing Die Wax, to ensure there are no stuck cases in his reloading press.

So now your die and shellholder are set up and properly positioned and the cases are lubricated. Insert the first case into the shellholder, seeing that it is pushed all the way into the shellholder's recess cut (and is, therefore, centered). Raise the ram through a full stroke of the press' arm and resize the case. Check to see that the decapping pin has removed the primer completely. If it has not, a bit more adjustment of the die in the press is in order, and then you'll repeat the process, as running the case through the die again will do no harm to it at this point.

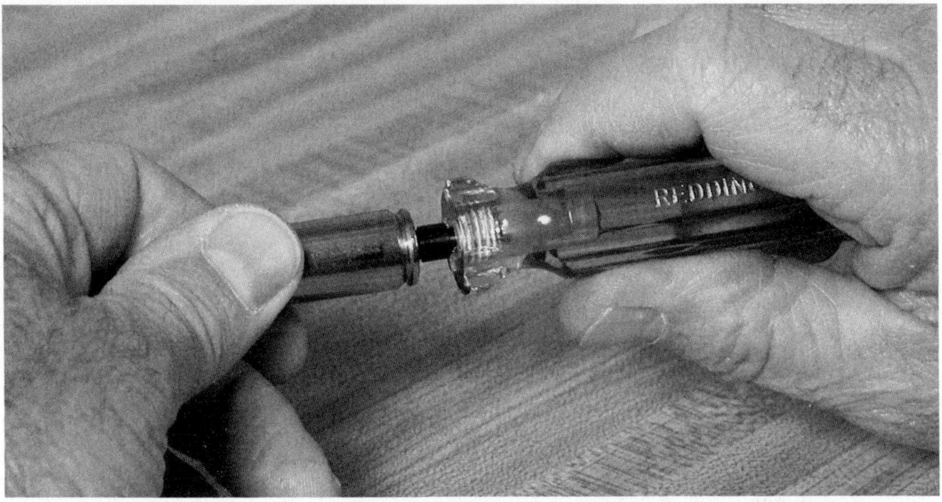

There are special primer pocket cleaning tools available from several companies, but a flat-blade screwdriver of the right size will also do the job.

You'll likely be inclined, at this stage, to reach for your priming tool and powder, but there's a couple cleaning jobs left to do before you get to those steps. First you must wipe off the case lube you used in the resizing process. A paper towel works great with the imperial Wax, and one sheet can be used on many cases before a new one becomes necessary. With other types of case lube, refer to the instructions for cleaning specific to the product.

The final step in the case cleaning could not be done until the primer was removed (via the decapping pin the sizing die), and that is the cleaning of the primer pocket. In this task, you must remove any residual primer residue and carbon. There are special tools made for this, but a properly sized flat-blade screwdriver tip will work just as well. Simply rotate the tip or blade to act as a rotary scraper, removing all matter of non-case material from the interior surface of the primer pocket.

CASE MEASURING AND TRIMMING

Finally, we have a case, which can be reloaded. Or do we? We spoke earlier of how a case grows in length and how brass flows. Now we must measure our resized brass to be sure it is within the specified allowable tolerances for overall case length. This is easiest done with a of good-quality dial caliper that has a resolution of at least one one-thousandth (0.001) of an inch. Many makers and resellers of reloading equipment stock this item, as do tool shops and even auto parts stores. Analog or

Measuring the overall case length of a resized case is necessary, so that you may know how much to trim.

digital versions are available and can be as reasonable as $25 to $30. I've even found a set of electronic calipers at Harbor Freight discounters on a sale for $9.95 that do the job! Just be sure the batteries are good on the fancier units.

The measurement for overall case length is taken from the flat of the case head to the opening of the case mouth. Remember to refer to the reloading manual for the correct data. The length is always expressed as a maximum dimension less ten-thousandths of an inch.(0.010-inch). Be sure you measure across the opening of the case mouth at its center line for the most accurate reading. As long as the case in question is within the safe tolerances, you can proceed with reloading the case. If not, the case must be trimmed.

The simplest and likely the oldest way to get the job of case trimming done is with the trim die. This type of die is machined to a minimum case length when measured against the case in the shellholder and the ram holding the shellholder hard against

A trim die.

A Redding Model 2400 Case Trimmer.

the bottom of the die. With the case trimming die thus in place, the user takes a fine mill file to the over-long brass and simply files the case back to the hardened flat top of the die. This is easy and leaves a good case with a square mouth. The only thing left to do after the filing is to deburr the inside and outside of the neck with a chamfering tool. This removes rough imperfections in the metal and also creates a slight taper on the inside of the case mouth to make bullet seating a bit easier when you reach that point.

The next level of trimming is accomplished with a case trimmer tool. There are many from a wide variety of manufacturers, and they can be either powered or hand cranked. The job of any trimmer is to hold a case so that a cutter on a shaft can precisely and squarely trim the case back to a pre-determined length set by the user, and do so in a repeatable manner. Some trimmers turn the shaft. Redding builds its trimmer as a true lathe, turning the case against the cutter, ensuring a square cut each time.

My personal favorite, when it comes to case trimmers, is the Redding Model 2400. It is not a "basic" machine, instead having a very advanced feature, a micrometer on the end of the cutter shaft that makes it perhaps the easiest to set up in comparison to other types. In the case of this machine, a primary cut is needed simply to square the case mouth. You'll then take an overall case length measurement using the caliper and simply turn the micrometer as many thousandths as necessary to gain the correct case length.

6 Bottleneck Case Priming

Most presses today come with some type of an integrated priming device. Normally, this is an arm attached to the press that will swing into position into a recess under the portion of the ram where the shellholder is held. These arms usually have provisions for changing from Small to Large primers. They may require some setup and adjustment, which will be described in your owner's manual for your particular the press.

When you begin to prime, safety is a major consideration. Safety glasses must be worn. The primer is designed to be sensitive to impact or crushing, so the improper seating of a new primer can cause it to ignite. A good set of safety glasses will remove the greatest potential danger to your physical well-being.

You will determine what primer is needed for the loading by consulting the loading data for your cartridge and specific load. Though I spoke of this before, for the sake of safety, I will stress that primers from different manufacturers have different levels of *brisance*. You should only use the primer brand and size specified for the

Priming the case. The arrow points to the the primer anvil, properly facing up.

Solidly bottom the primer in the primer pocket. This is something you'll learn to feel as you reload more, but it never hurts to do a visual inspection after this operation, to make sure the primer is seated neither too deeply nor not enough. The newly seated primer should be flush with the head of the case.

Priming can be done in a variety of ways. This hand-priming tool from Lee helps ensure minimal handling of the primers.

load you are using. At this time, you should give one last inspection to the case, to be sure there is no primer crimp remaining and to see that there is no obstruction in the flash hole.

To prime a piece of brass, place a clean, sized, and de-primed case in the shellholder and raise the ram about a third of its travel, or enough to have the primer arm, with its cup, freely rotate into position. Remove a primer from its packaging. Clean hands are important, as is a clean working environment; things like case lube can compromise the integrity and operation of a primer if the lube contaminates the priming compound. Place the primer into the "cup" of the priming arm with the primer's anvil side up and its smooth side down. Rotate the arm so the primer is within the slot in the body of the press ram and then lower the ram until you feel the primer seat to a hard bottom. Do this gently and smoothly to be certain the primer isn't being crushed. If the primer does not fully seat, refer to the specific instructions for your press and make the needed adjustment. If the primer protrudes so far that the case cannot be removed, it is fine to leave the case in place while the seating depth is adjusted. Once the primer is fully seated, be sure to inspect your work and see that the primer is indeed is flush with the case head.

In addition to the priming arm available for most single-stage and turret press operations, many manufacturers provide a variety of dedicated priming devices. Some bench-mount, some are handheld, and others provide an upgrade on the press itself.

I have two favorites, each of which can be considered basic in their application. First is a Redding Straight Line Priming System, which fits on the Redding Big Boss, Big Boss II, and T-7 Turret presses. It is quite fast and easy to use, and it comes with all that is needed to work with both Large and Small primers. One unique feature it has is the safety shield. This is a blued steel barrel cover that screws onto the base of the priming system and encases the aluminum primer pick-up tube. If an accident were to happen, the resulting discharge will go straight up into the ceiling instead of impacting the press operator. This is well designed and appreciated by serious handloaders who wish for added safety with volume reloading practices.

My other long-time favorite is the Lee Hand primer. Handheld with a built-in primer "flipper," this tool is inexpensive and works extremely well. It comes with all that's needed to seat both Large and Small primers and has great "feel" on seating. The only negative is that one must buy special shellholders in order to use the tool with different calibers. These are not expensive, but an added requirement nonetheless.

Regardless which method you choose, with the primer now fully seated, be sure to inspect your work and see that it is flush with the case head. That done you are ready for the next steps of powder measurement and filling the case with the powder.

7

Bottleneck Case Powder Measurement and Pouring

The measurement of powder is a critical item in your loading. Precision here yields safety and consistency to the loads you create.

The first and foremost item you'll need is a good scale, one that is consistent and capable to resolving a single tenth of a grain. "Grains" are the unit of measurement for smokeless powder, and there are 7,000 grains in one pound. A powder scale must be accurate to $^{1}/_{10}$-grain, which equates to $^{1}/_{70,000}$-pound.

Scales come in all forms and at all price points. There are traditional beam scales, electronic scales, and even measuring systems where you preset the desired load and the device dispenses it until the correct weight is in the weighing pan. We will focus on the beam scale, as it is accurate, inexpensive, and not affected by changes in atmospheric pressure, as electronic scales are known to be.

A Redding Model 2 Beam Scale with magnetic dampening.

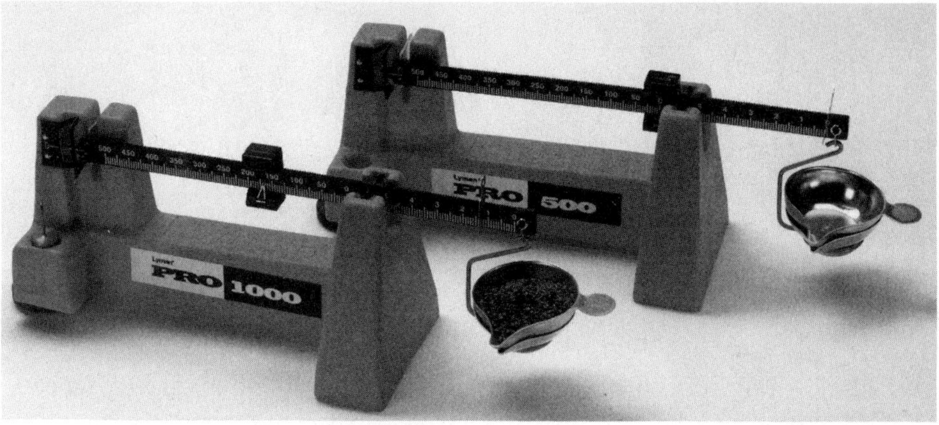

Lyman's Pro Series scales are available in two different weight capacities.

Your scale should be placed in an area without air movement, which can impact accuracy. Fans, windows, and AC vents are the enemy here. The beam scale is assembled by setting the base, mounting the beam on its bearing surfaces, and attaching the pan hook and pan. This done, you need to calibrate the scale to zero. This is generally done by the turning of a screw in the base to level the "0" mark with the pointer on the end of the beam. Once leveled and zeroed, the scale is ready for use.

Beam scales have changed very little over the years, other than the inclusion of dampening devices to more quickly "settle" the beam on the final weight. To settle the scale means that the beam's swing above and below the final weight stops, allowing you to determine the given weight of the charge in the pan. Methods of dampening include oil, whereby a small paddle protrudes from the beam into a well of oil that slows the beam's movement, and magnetic, whereby a vane on the beam rides between two magnets, which retard the beam's speed of movement. A dampened scale is no more accurate by nature, but it does speed the process of the scale's settling, thereby speeding the final measurement of powder.

In addition to your scale, you will need a plastic spoon, a powder trickler, and a powder funnel. You'll use a plastic spoon to eliminate static electricity, which can cause the powder to stick to the spoon, and also to remove the odd chance of any spark from the same. You'll derive the type and weight of the powder charge from a powder loading data manual or from the tables online at various powder company websites. I ask that you use only sources considered official by the powder company and not something posted on a blog or chat board. Official sources

The poises on the scale beam.

are listed to perform within SAAMI specs and have been verified in a SAAMI spec pressure gun; the pressure signs we may read about on blogs and chat boards may be very misleading. As a beginner and a safe reloader, know that pressure data derived from an industry standard crusher or piezo electric pressure gun in the lab of a powder or ammunition company is where you want your personal safety and the safety of those around you at the range to be assured.

Now, let's talk about weighing your powder charge The beam scale generally has two moveable weights, known as "poises." One is large and generally on the opposite side of the balance point from the pan and pan hook. The other is lightweight and is on the same side as the pan hook. On most scales, the heavy poise is calibrated to move in five grain increments, usually from five to 250 or 500 grains. The smaller poise on the pan side is calibrated in 1/10-grain increments from zero to five grains. Using these two poises, you'll preset the desired charge of powder the loading manual calls for on your particular load. If that charge is 26.2 grains, move the large poise to the 25 grain setting and the small one to one grain plus two-tenths.

Once your scale is set to your desired weight, begin to very slowly add powder to the pan, trickling it off the spoon slowly and carefully. As you approach the desired weight, the pointer on the scale will begin to move up toward the matching line of the zero mark. (On some beam scales, like the Redding Model 2, there are also graduations on the pointer end indicating plus and minus 5/10-grain in 1/10-grain increments.) Once you see movement of the pointer toward the zero mark, you should change over to your powder trickler.

The powder trickler is a tool unique to reloading. It is comprised of a funnel-like device that has a tube inserted in a cross-drilled hole near its base. Generally, there's a stand of some sort incorporated, which allows the tube to be at the proper height to drop powder into the pan of the scale. Inside the tube is a screw thread and, within the base of the

Hornady makes a combined trickler and electronic scale called the Auto Charge Powder Measure.

funnel, the tube has a small entry hole for the powder. One end is open, while the other usually has a small knurled knob that is easily turned by hand.

As you turn the knob clockwise, the individual grains or flakes of powder (not to be confused with the grain as a measurement of weight), travel in the screw thread to the open end. This allows you great precision for adding small amounts of powder to the pan as you continue to rotate the knob. Slowly add powder until the pointer on the beam meets the center mark on the frame.

The powder is in the pan of the scale and now needs to be transferred to the primed case. For this you'll use a powder funnel. Another item specific to reloading, the powder funnel is generally of some nonconductive, anti-static material for easy flowing of all the powder into the case as it's poured from the pan. In such a funnel, there is a large funnel on the top, but the outlet tube is large enough to fit over a number of cases up to, normally, .45-caliber.

The powder trickler is used to add the final, incremental amount of powder to zero the scale to create the correct load.

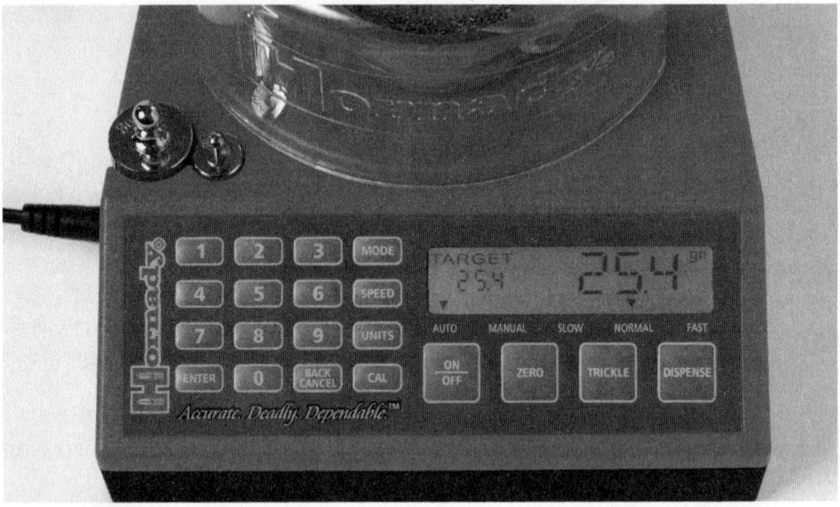

The electronic scale from Hornady is very precise. Just remember that electronic scales can be sensitive to atmospheric pressures, and that some adjustments to such a scale's calibration may be necessary.

Within this end is a reverse taper, which allows the case to enter the tube until the case mouth meets solidly against the internal cone shape, creating a seal so that all the powder goes in the case. Powder funnels generally have a small through-hole to accommodate cases as small as .20- or .22-caliber, and adapters are available to take this as small as .17-caliber.

Holding the pan from the scale in one hand, and the funnel over the case in the other hand, slowly pour the powder into the funnel. I often pour it in a circular pattern about midway up the funnel, swirling it as I pour. I find this technique allows the powder to find its best way to settle in the case. This is important with powders that may nearly fill the case at a given charge weight. Trust me, a little extra time and effort here pays off with easier bullet seating, which we'll get to next.

8 Bottleneck Case Bullet Seating and Crimping

With the powder phase now done, you can move to bullet seating and crimping.

The seating die is now needed in your press. It's important to recognize that the seating die does not impact case dimensions in any way (except to roll-crimp the bullet into the case mouth on some, but not all, calibers). What the seating die is designed to do is align the case and hold the seating stem in a concentric manner in order to seat your

Inserting the seating die into the press.

bullet in the best possible fashion. As this is a basics book, I will focus on the standard type of seating die in this discussion.

To set up the standard seating die, we must first screw it into the press with the appropriate shellholder installed. With the ram completely raised, screw the die down until it touches the shellholder. Remember, dies are designed to use the interface of the die base and the top of the shellholder at set, specific dimensions. Now, screw the die back out two full turns to remove the crimp, if any, from the setup equation (though at this point and this point only). With a 14-turns-in-one-inch thread pitch, or 14 TPI, this action moves the die up approximately 0.144-inch. This means the internal crimp ring of the die is well above where the case neck will be at the top of ram stroke. Now, lock the die lock ring to maintain this position; there's no need to lock the set screw just yet.

Seating the bullet. Hold and guide the bullet into the die, but be careful not to pinch your fingers!

Place a primed and powder-loaded case into the shellholder and set the bullet in the case mouth; here you will see where the benefit of neck chamfering comes into play. Push the bullet into the case a bit, though the neck tension should keep you from moving the bullet in very far. You just want to set the bullet in a proper attitude so that, when it enters the die to contact the seating plug, it is as straight as possible.

You're almost ready to seat the bullet, but, before pulling the handle to raise the ram and do just that, you must back the seat plug out so you can soft-seat the bullet. Soft-seating allows you to get the bullet started, as well as centered and secured by neck tension, so that you can then make the needed adjustments to gain the proper overall *cartridge* length (I talked about overall *case* length before, which, as its name implies, is a measure of just the brass; overall *cartridge* length includes the seated bullet). This is another measurement that will be found in your powder loading data manual. Cartridge overall length, or "COL," is important to safe functioning of your reloads in your firearm. Again, do not substitute. Aim to achieve the proper COL for the bullet, case, primer, and powder specified in the data manual. This will ensure safety, accuracy, and proper function.

Using your dial or electronic caliper, measure the case from the bullet tip, known as the "meplat," to the flat of the case head where the primer is located. This will give you your COL measurement of the loaded round after soft-seating. That done, now you must adjust the seating plug to push the bullet just deep enough into the case to provide the proper COL dictated by the loading data. The thread pitch

Measuring COL, the cartridge overall length, with calipers.

The completed round ready to use!

of the seat plug is the determining factor in how far you'll screw in the seating plug to gain the desired length. On my Redding seating dies, for instance, the thread pitch is 20 turns per inch, or 20 TPI, which conveniently gives me 0.050-inch per rotation.

Let's look at an example. Say my powder data manual specifies 2.224 inches as a maximum COL. My initial soft-seat has left me with a loaded round of 2.361 inches. Knowing the pitch of the thread is 20 TPI and the rotation will yield 0.050-inch of travel, I will start by turning the seating plug two complete rotations. As there are no markings on the seating plug, I use the number stamped in the top as my reference. If I was perfect—and no one ever is—I should get a measurement of 2.261 inches. However, reality says I overturned slightly and I am at 2.248 inches. I know must slowly and incrementally turn the seating plug in to reach my goal of 2.224 inches. In other words, I need to shorten the loaded round by 0.026-inch, which is ever so slightly more than a half-turn.

Do not try and hit this on one try! You will likely overseat. You're going to have to use the trial and error method. I will go just over a quarter-turn, check, and then begin a series of very slight bumps of the rotation to get the length I need, measuring each time I make one of those small adjustments until I have reached the goal of my COL at exactly 2.224 inches.

One thing to take note of is that some bullets have a "cannelure," or "crimp ring," and, if you wish to crimp, you can begin that process now. This is why you did not earlier lock into place the major die lock ring.

A "roll crimp" is exactly as the name implies. You'll use a raised metal portion of the die's internal design called a "crimp ring" to roll the mouth of the case into the cannelure or crimp ring embossed on the bullet. If you've done everything right up to this point and your nearly completed round is of the proper length, the cannelure of your bullet should be aligned at the case mouth.

To set the crimp, first loosen the major die lock ring and screw it up and out of play for a moment. You'll also need to back out the seating plug you worked so hard to adjust and remove it from the operation. Next, raise the ram (with the loaded round in the shellholder) to the uppermost position. While the case is in this position, begin screwing the die down until you feel contact. This is the crimp ring in the die contacting the case mouth. Now, back off the ram and screw the die in another eighth of a turn (with a 14 TPI of the die body, this equates to about 0.010-inch). This should be enough to provide a good crimp. Lock the major die body lock ring and crimp the case. Inspect to see the crimp has sufficiently rolled the neck in, and now, with a completed cartridge in the shellholder and the ram fully up, slowly and carefully screw the seating plug back in until you feel contact on the bullet ogive. Lock the seating plug in position and lock the set screw in the major die lock ring. Sounds like a lot? Well, the good thing about this entire process is that you have now made all the adjustments necessary to seat your bullets to proper length and crimp them all in one step!

SECTION C:
Reloading the Straight-Wall Case

Straight-Wall Case Dies, Case Lubrication and Case Trimming

While much of this is a rehash of the section on bottleneck cases, the differences to the process cannot be properly handled by a simple addendum. Therefore, the following chapters in this section form a complete how-to on the straight-walled case loading process.

THE SIZING DIE

As you know from reading the initial chapters in this book, the firing of the cartridge expands the brass, and so the first chore must be to return the case to a reusable state. The sizing die is the tool that performs this task. It is caliber-specific and needs to be fitted into an appropriate reloading press.

You'll begin with a steel sizing die, one that resizes the entire body of the case. To do this, the die must be properly set up in the press with the appropriate shellholder. The shellholder is also specific, though not to caliber, rather to the dimensions of the case head. Therefore families of cases can use the same shellholder. A family of cases may be one like that of the basic .30-06 case, which will also work for the .25-06 Rem., .270 Win., .280 Rem., and the .35 Whelen, as well as wildcat cartridges using the .30-06 case head dimensions.

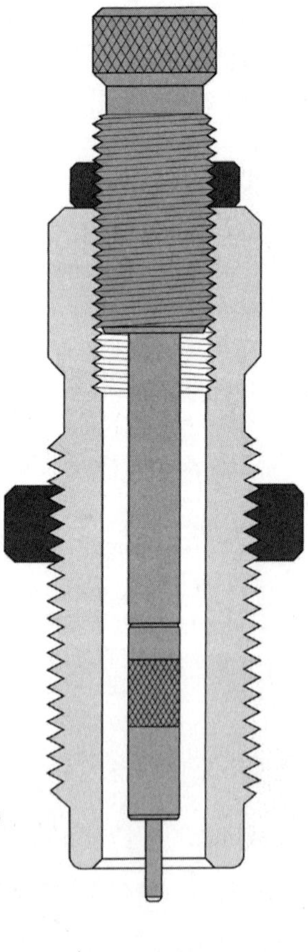

A steel straight-wall case sizing die cutaway showing the decapping assembly and lock rings.

A variety of shellholders in a storage case.

With the proper steel sizing die and the correct shellholder, in hand, you next need a process for setting up the die in the press. Again we will focus on the basic setup, which will allow us the greatest possible sizing, assuring good bullet tension and a smooth operation of the finished cartridge in any chamber that falls within SAAMI specs.

Setup of your sizing die begins by inserting the shellholder into the top of the press' ram. The shellholder is usually held in place by some type of spring clip, which will act to both align and retain the shellholder in its proper position. Now, raise the ram to its topmost position and move the major die body lock ring upward and out of the way, so that the die can be screwed into the press. Now screw the die down until it makes a complete contact with the top of the shellholder.

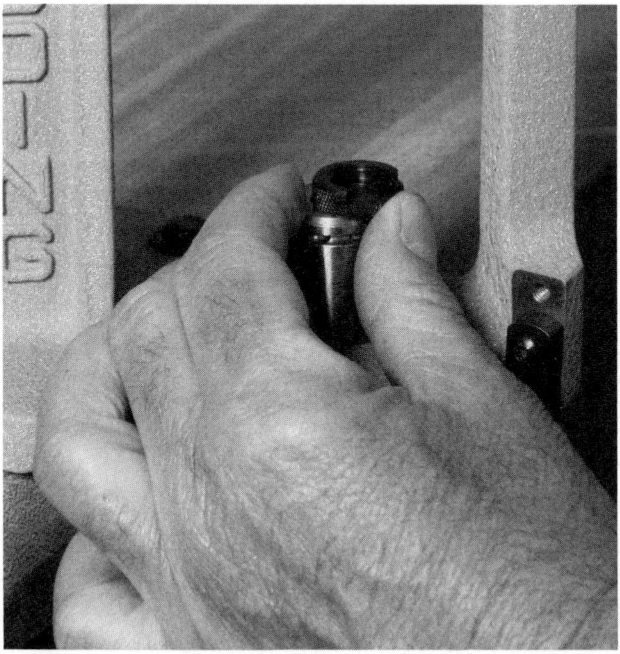

Inserting the shellholder into the press' ram.

At this point, you must back off the ram slightly (lower it a bit) and screw the die in an additional eighth of a turn. This creates the situation known as "cam-over" on a traditional reloading presses. This slight cam-over is needed to assure that the entire case is fully sized for proper chambering. Once this task is complete, lower the ram to its original position.

Before you size the first case, you must adjust the depth of the decapping pin's travel and, therefore, the length of its rod. This is critical to preventing the decapping pin retainer from bottoming out and bending the rod. The goal with decapping is to use only enough pin protrusion to completely pop the primer from the cup in the case head, and not to overdo this by crashing the expander into the web of the case and causing damage to it.

Though dependent on the brand of die set you're using, adjusting the depth of the decapping pin travel is generally done by loosening the lock nut and rotating the rod out, or loosening the collet and withdrawing the rod upward so that the decapping pin only slightly protrudes below the bottom of the die.

We are almost ready to size the first case, but, with a steel die and a straight-wall cartridge case, you must lubricate the brass before sizing. Case lubrication is critical to keeping from the die from sticking on and ruining a case. There are a large number of lube types and lube systems in the marketplace, and every reloader has their favorite.

As the steel sizing die is truly a full-length die and is, therefore, in contact with the full length of the case. On a straight-wall case, this creates almost complete surface contact on all parts of the case exterior under, literally, thousands of pounds of pressure that compress and push the brass into the desired dimensional shape required for chambering. Here again, we can study the word "obturation." As we compress and reshape the brass in the resizing stage of reloading, that brass wants to expand again slightly (obturate) after

Inserting the full-length steel sizing die for straight-wall cases.

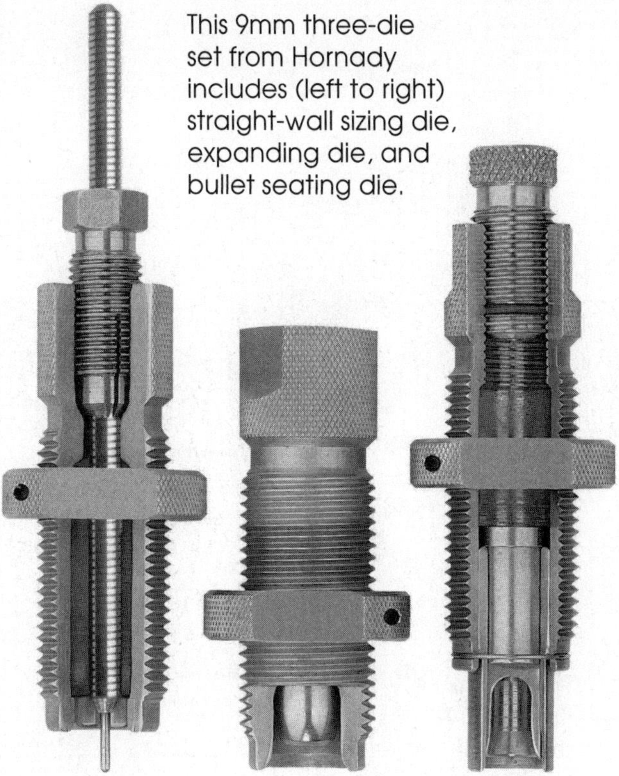

This 9mm three-die set from Hornady includes (left to right) straight-wall sizing die, expanding die, and bullet seating die.

the force of the die on the case is released. This obturation presses the entire length of the case against the interior walls of the die, and lack of proper lubrication will keep the case from being withdrawn from the die as you push the handle in the direction to lower the ram. The compound leverage of the ram has great strength in both directions, and a case can be so stuck so severely that the rim of the case can be torn off in the process.

My favorite lube is Imperial Sizing Die Wax. It is very easy to use and control. Imperial comes in a tin like shoe polish. To use it, simply rub the flat of your index finger tip on the surface of the wax until a small buildup occurs. Then, rub your index finger and thumb together, gaining an even coating on the surface of each. Beginning with the case head end, roll the case between your finger and thumb from the case head to its mouth. Cases tend to stick at the bottom 20 percent, as this is where the brass is the thickest. I like the Imperial product, because you almost instinctively hold the case by its mouth with the hand that does not have the lube on it. Therefore, the process of getting the greater coating of lube where it is needed most becomes automatic.

Imperial Sizing Die Wax.

Many people like lube pad products, and I have had good experience with them. I am personally adverse to the spray lubes. While they look easy to use, I immediately find a number of issues with them. Most require a waiting period of 30 minutes before loading. I am not a patient person, but not waiting causes stuck cases. Additionally, as we usually rank and file cases and spray them from the top, the cases don't always get the necessary amount of lubricant in the critical lower area.

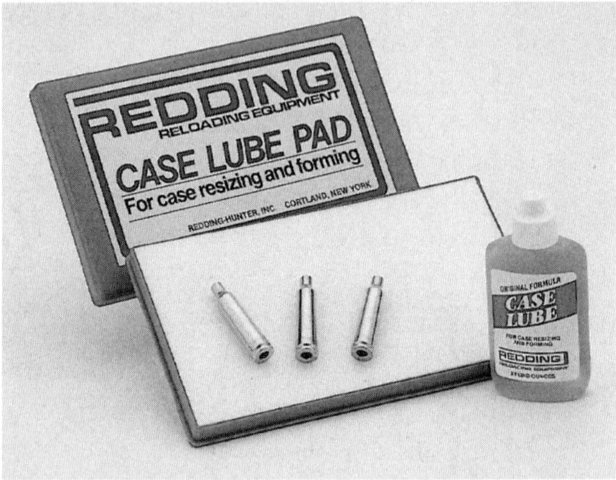

The case lube pad from Redding can produce good results.

So our die and shellholder are now set up and properly positioned, the case is lubricated, and now we insert the case into the shellholder, seeing that it is pushed all the way into the shellholder's recess cut and is, therefore, on center. Raise the ram through a full stroke and resize the case, then check to see that the decapping pin has removed the primer completely. If it has not, a bit more adjustment is in order. If you needed to adjust your decapping pin, repeat the resizing process, as running the case through the die again will do it no harm.

With the primer cleared, you are almost ready to begin your reload, but before you reach for the powder, you have a couple cleaning jobs to do. First, you must wipe off the case lube used in sizing. A paper towel works great, if you're using the Imperial product like I do. With other types of case lube, refer to the instructions for cleaning that came with the product.

The final step in cleaning the case could not be done until the primer was removed, and that's the cleaning of the primer pocket to remove any residual primer residue and carbon. There are special tools made for this, but a properly sized flat-bladed screwdriver tip will work just as well. With the screwdriver, you'll rotate the tip or blade to act as a rotary scraper, removing all manner of non-case material from the interior surface of the primer pocket.

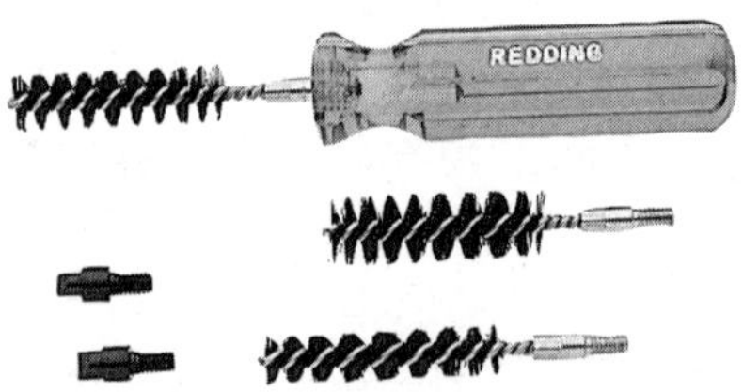

A case cleaning kit with brushes and primer pocket cleaners for large and small primer pockets.

CARBIDE SIZING AND EXPANDING DIES

So far I've talked about using steel dies for your reloading of both bottleneck and straight-wall cases, but a carbide sizing die can be used on a straight-wall case. The term "carbide die" is somewhat an overstatement, as it's really a steel die holding a carbide ring near its base to do the sizing. The carbide ring is ground to an internal dimension specification that will size the length of the case to a dimension to fit your gun's chamber, but, more importantly, it's also a dimension small enough to hold the bullet tightly upon seating.

The set up and use of the carbide die is generally the same as it is for the steel die, but with two notable exceptions. The carbide die should not be run with cam-over, as sometimes the carbide ring is very close to the bottom of the die. Carbide is quite brittle, and over-camming could cause the ring to crack under a compressive loading from cam-over of the press. So, with a carbide die, we will simply run the die body down until it makes contact with the shellholder and stop. Second, the carbide surface is very hard and slick. As such, it does not require the use of case lube. This is also a function of the narrow contact surface of the ring at any given time against the case, which greatly lessens the surface contact throughout the sizing process.

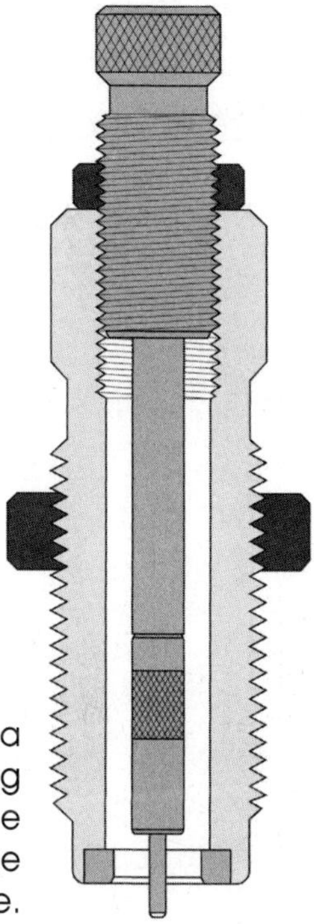

A cutaway of a carbide sizing die showing the carbide ring at the bottom of the die.

STRAIGHT-WALL CASE EXPANDING

All the basic steps involved in loading the bottleneck case are the same for the straight-wall case. However, the bottleneck case has a body that is larger in diameter than its neck, and that allows an expander to be placed on the decapping rod stem of the sizing die. In a straight-wall case, this isn't possible, so we must expand the case in a third die called, sensibly enough, an "expanding die." Some makers use a true expander, while others have a die to simply bell the case mouth and allow the bullet to enter more easily, especially flat-based pistol bullets.

An expander die has a true expander within it. This is well illustrated with the Redding die, which has a long expander "plug" containing numerous machined surfaces that provide specific benefits to the reloading process. You've learned that, when sizing an outside diameter (OD), you create an inside diameter (ID) that is dependent on the wall thickness of the brass. You also learned that neck tension and start pressures are important to safety, quality, and consistency of your reloaded ammunition. Therefore, proper neck expanding in a straight-walled case is an important factor in its reloading.

While this is a book on basic reloading, quality of the finished product and a good level of performance are certainly part of those basics. A close look at the Redding expander in the numbered diagram shows some important design considerations that have a positive impact on the finished round. First (1) you'll see a taper at the nose of the plug, which allows for its easy entry into the case while the mouth is being opened. Next you'll see a long parallel section (2), which is a specific dimension less than that of the bullet diameter—this is the true expander portion of the plug. This section is as long as the section of the bullet that will be contained within the case; it is here that you'll set your neck tension to provide a true and smooth mating surface by removing any irregularities in the brass away from the bullet interface surface. Next, up you'll see a small square lip (3). This is to create a "seat" for the bullet to sit in, something especially important for a flat-based bullet. Finally, at the top you'll see a flared section (4), which, likewise, flares the case mouth for easy bullet entry. That's an exhaustive explanation, but this

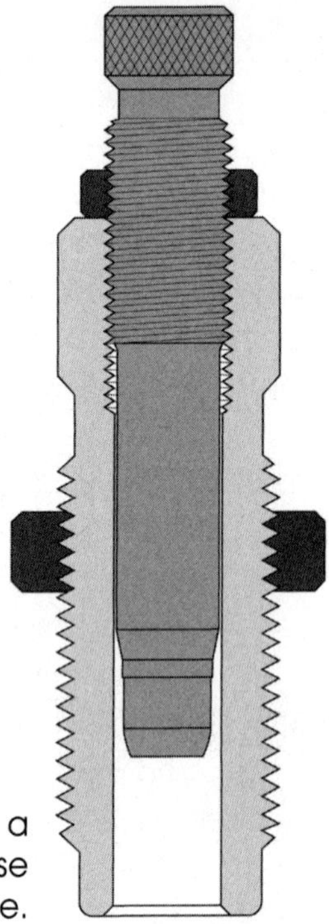

Cutaway of a straight-wall case expander die.

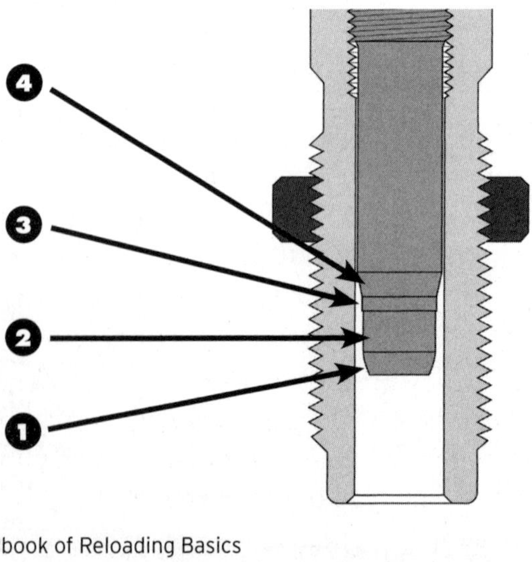

Beginning to see the case mouth flare, or "bell."

Note the now completed flare, or "bell," on the case mouth.

kind of expander will have cared for the case and your loading needs well beyond the basics.

To set up the expander die, screw the die into the press with the shellholder in place. Depth is not as critical here, as the die body only acts as a tool to hold the expander within it so that it can do its job. With a sized case in the shellholder and the ram in the full up position, screw the expander plug into the case until you begin to feel firm resistance. This is the expander taking the case back to the proper inside dimension.

Back off the ram and again screw in the expander, deeper this time. You'll be looking to see, after each operation of the press, how far the expander has penetrated the case. This is a test-and-check program, but well worth the end result. You'll know you have properly set the die and expander plug, when the case prints with the expanded section and you can see the slight flare of the case mouth.

TRIMMING STRAIGHT-WALL CASES

Remember that I spoke earlier about how a case grows in length and how brass flows when a cartridge is fired. These things also hold true when you're resizing your brass, so now you must measure to be sure that the brass after sizing is within the specified allowable tolerances for case length. This is easiest done with a set of good quality dial calipers that have a resolution of at least one one-thousandth (0.001) of an inch. Many makers and resellers of reloading equipment stock this item, as do tool shops and even auto parts stores. Analog or digital versions are also available and can be as reasonable as $25 to $30. I found a set of electronics at Harbor Freight discounters on a sale for $9.95 that do the job. Just be sure that batteries are good on such units before using.

The case length measurement is taken from the flat of the case head to the opening of the case mouth. You will refer to the reloading manual for this data. The length is always expressed as a maximum dimension less ten thousandths of an inch (0.010-inch). Be sure you measure across the opening of the case mouth at its center, for the most accurate reading. As long as the case in question is within the safe tolerance recommended by the loading manual, you can proceed with the loading process. If not, the case must be trimmed to be safely used in your firearm.

Of all the case trimming tools and methods, the simplest and likely the oldest is the trim die. This type of die is machined to a minimum case length when the case is in the shellholder and the ram is holding that shellholder hard against the bottom of the trim die. The user then takes a fine mill file and simply files the case back to be even with the hardened flat top of the die. This is easy, direct, and produces a good case with a square mouth, when properly done.

After any trimming operation, you must always deburr the inside and outside of the neck with a chamfering tool. This removes rough burrs and also creates a slight taper on the inside of the case mouth to make bullet seating a bit easier when you reach that point.

As an alternate, and as the next level up in trimming methods, you can use a case trimmer. There are many available from a wide variety of manufacturers and they

A trim die.

can be powered or hand cranked. The basis of any trimmer is to hold a case so that a cutter on a shaft can precisely and squarely trim the case back to a predetermined length set by the user. Some trimmers turn the shaft. Redding builds its trimmer as a true lathe, turning the case against the cutter and assuring a square cut each time.

My favorite case trimmer is the Redding Model 2400, which has the advanced feature of a micrometer on the end of the cutter shaft, making it, perhaps, the easiest to set up in comparison to other types. In the case of this machine, a primary cut is needed simply to square the case mouth. Next you'll take a length measurement using the caliper and simply turn the micrometer as many thousandths as indicated to be necessary to gain the correct case length.

10 Straight-Wall Case Priming

Most presses today come with some type of an integrated priming device. Normally, this is an arm attached to the press that will swing into position into a recess under the portion of the ram where the shellholder is held. These priming devices usually have provisions that allow change-out from Small to Large primers.

When we begin to prime, safety is a major consideration. Safety glasses must be worn. The primer is designed to be sensitive to impact or crushing, so the improper seating of a new primer can cause a primer to ignite. A good set of safety glasses will remove the greatest potential danger to your. I will also stress again that, for the sake of safety, primers from different manufacturers have different levels of *brisance*: you should only use the primer brand and size specified for the load you are using. Never swamp one brand of primer for another, nor another size for another. Finally, give one last inspection to the case,

Priming the case. Note the primer's anvil facing up.

to be sure there is no crimp remaining and to verify that there is no obstruction in the flash hole.

To prime a freshly sized and properly trimmed straight-wall case, place your case in the shellholder and raise the ram about a third of its travel, or enough to have the primer arm freely rotate into position. Remove a primer from its packaging. Clean hands are important here, as is a clean working environment; things like case lube can compromise the integrity and operation of a primer if that lube contaminates the priming compound. Place the primer into the cup of the priming arm with the primer's anvil side up, its smooth side down. Rotate the arm so the primer is within the slot in the body of the press ram, then lower the ram to feel the primer seat to a hard bottom. Do this smoothly and gently, to be certain the primer isn't crushed. If the primer seats hard or does not fully seat, refer to the specific instructions for your press and make the needed adjustment. If the primer protrudes so far that the case cannot be removed, it is fine to leave the case in place while the seating depth is adjusted.

With the primer fully seated, be sure to inspect your work and see that the primer is flush with the case head. Inspection done, you're ready for the next steps of powder measurement and filling the case.

ALTERNATE PRIMING METHODS

Many manufacturers provide a variety of dedicated priming devices. Some bench mount, some are handheld, and others provide an upgrade to the press itself. I have two favorites. First is a Redding Straight Line Priming System, which fits on the Redding Big Boss, Big Boss II, and T-7 Turret presses. It is quite fast and easy to use, and it comes with all that's needed to work with both Large and Small primers. One unique feature is the safety shield. This is a blued steel barrel cover that screws onto the base of the priming system and encases the aluminum primer pick-up tube. If an accident were to happen, the resulting discharge will go straight up into the ceiling, instead of impacting the operator. This is well designed and appreciated by serious handloaders who wish for added safety with volume reloading regimens.

My other long time favorite is the Lee Hand Primer. Handheld and with a built-in primer "flipper," this tool is inexpensive and works extremely well. It comes with all that's needed to seat both Large and Small primers and has great "feel" on seating. The only negative is that one must buy special shellholders to use with different calibers. These aren't expensive, but, nonetheless, they are still an added requirement.

11 Straight-Wall Case Powder Measurement and Pouring

Precision powder measuring yields safety and consistency to the loads you create. The first and foremost item to getting precise powder measurements is a good scale, one that is consistent and capable to resolving one-tenth of a grain. Grains are the unit of measurement for smokeless powder. There are 7,000 grains in one pound. A powder scale must be accurate to $^{1}/_{10}$-grain, which equates to $^{1}/_{70,000}$-pound. In addition to the scale, other powder tools you will also need are a plastic spoon, a non-conductive powder funnel, and a powder trickler.

Scales come in all forms and all price points. There are traditional beam scales, electronic scales, and even measuring systems where you preset the desired load and the device dispenses it until the correct weight is in the pan. I'll focus on the beam scale, as it is accurate, inexpensive, and not affected by changes in atmospheric pressure, as electronic scales are known to be. Your scale should be placed in an area without air movement, which can impact accuracy. Fans, windows, and AC vents are the enemy here.

The beam scale is assembled by setting the base, mounting the beam on its bearing surfaces, and attaching the pan hook and pan. This done, you next need to calibrate the scale to zero. This is generally done by the turning of a screw in the base to level the "0" mark on the scale's frame with the pointer on the end of the beam. Once leveled and zeroed, the scale is ready for use.

Beam scales over the years have changed very little, other than the inclusion of dampening devices to more quickly "settle" the beam on the final weight. To "settle" the scale means that the beam's swing above and below the final weight stops, allowing you to accurately derive the given weight of the charge in the pan. Two methods of dampening include oil, where a small paddle protrudes from the beam

Level the scale to obtain a zero, then use the weights to set your desired charge down to the tenth of the grain called for in your loading data.

into a well of oil, slowing the beam's movement, and magnetic, where a vane on the beam rides between two magnets, again, retarding the beam's speed of movement. A dampened scale is no more accurate by nature, but it does speed the process of the scale's settling and, so, speeds reading the final measurement of powder.

In addition to the scale, you'll also need a plastic spoon, a powder trickler, and a powder funnel. You'll use a plastic spoon to eliminate static electricity, which can cause the powder to stick to the spoon and also to remove the odd chance of any spark. You'll decide on the type and weight of the powder charge from your powder loading data manual or from the tables online at the powder company's website. I ask that you only use sources considered official by the powder company and not something posted on a blog or chat board. Official sources are listed to perform within SAAMI specs and have been verified in a SAAMI spec pressure gun. The pressure sign you may read abut in blogs and on chat boards may be very misleading. As a beginner and a safe reloader, pressure data derived from an industry standard crusher or piezo electric pressure gun in the lab of a powder or ammunition company is where you want your personal safety and the safety of those around you at the range to be assured.

When the pointer on the scale moves to zero, the weight is correct. Now you're ready to move your powder charge to the case.

On to the weighing of the powder charge. The beam scale generally has two moveable weights, or "poises." One is large and generally on the opposite side of the balance point from the pan and pan hook. The other poise is lighter in weight and is on the same side as the pan hook. On most scales, the heavy poise is calibrated to move in five-grain increments in a range generally from five to 250 or 500 grains. The smaller poise on the pan side is calibrated in 1/10-grain increments from zero to five grains. Using these two poises, preset the desired charge of powder your reloading data manual calls for on the particular load you've chosen. If that charge is 10.2 grains, move the large poise to the 10-grain setting and the small one 2/10-grain.

Now you can begin to very slowly add powder to your pan, trickling it off the plastic spoon slowly and carefully. As you approach the desired weight, the pointer on the left side of the scale will begin to move up toward the matching line of the zero mark. (On some beam scales like the Redding Model 2, there are also graduations on the pointer that indicate plus and minus 5/10-grain in 1/10-grain increments.) Once you see movement of the pointer, you should change over to your powder trickler.

The powder trickler is a tool unique to reloading. It is comprised of a funnel-like device with a tube inserted in a cross-drilled hole near its base. It generally has a stand of some sort, allowing the tube to be at the proper height to drop powder into the pan of the scale. Inside the dispensing tube is a screw thread, and within the base of the funnel the tube has a small entry hole for the powder. One end is open, while the other usually has a small knurled knob that is easily turned by hand.

As you turn the knob on the trickler clockwise, the individual grains (not to be confused with the grain as a measurement of weight) or flakes of powder travel in the screw thread to the open end. This allows great precision in your powder loading, by enabling you to add very small amounts of powder to the pan as you continue to rotate the knob. Slowly, you'll add powder, until the pointer on the beam meets the center line on the frame. Now you have the precise amount of powder

Using the powder funnel makes adding the powder to the case simple and clean. The process with these simple tools is the same whether you're loading bottleneck or straight-wall cases.

specified by the placement of the poises, and that powder is ready to be transferred to the primed case.

For transferring the powder from the weighing pan to the primed case, you'll use a powder funnel. Again an item specific to reloading, the powder funnel is generally of some nonconductive, anti-static material for easy flowing of all the powder into the case as it's poured from the pan. The funnel's outlet tube is large enough to fit over a number of cases, normally up to .45-caliber. Within its end is a reverse taper, allowing the case to enter the tube until the case mouth meets solidly with the internal cone shape. This creates a seal so that all the powder goes in the case. Powder funnels generally have a small through-hole to accommodate cases as small as .20- or .22-caliber, and adapters are available to take this as small as .17-caliber.

Holding the pan from the scale on one hand, the funnel over the case in the other hand, slowly pour the powder into the funnel. I often pour it in a circular pattern from about midway up the funnel, swirling it as I pour. I find this technique allows the powder to find its best way to settle in the case. This is important with powders that may nearly fill the case at a given charge weight. A little extra time and effort here pays off with easier bullet seating later.

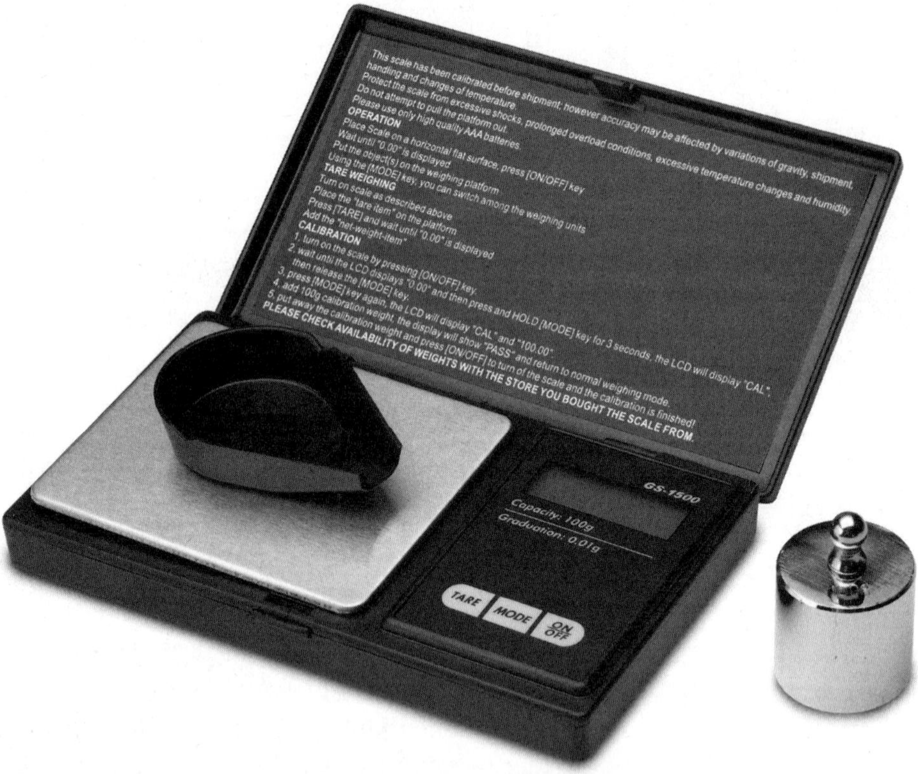

Electronic scales come in a huge variety. This is actually a pocket-sized digital scale from Hornady, an excellent tool for reloaders who don't have a lot of bench space, as well as for those using a portable reloading setup, such as a benchrest shooter might do.

12 Straight-Wall Case Bullet Seating and Crimping

With a properly trimmed and primed case loaded with powder, the seating die is now needed in your press. The seating die does not impact case dimensions in any way (except to roll crimp the bullet into the case mouth on some, but not all, calibers). The seating die is designed to align the case and hold the seating stem in a concentric manner so as to seat the bullet in the best possible fashion.

Inserting the straight-wall seating die into the press.

To set up the standard seating die, you must first screw it into the press with the appropriate shellholder installed. With the ram completely raised, screw the die down until it touches the shellholder. Remember, dies are designed to use the interface of the die base and the top of the shellholder to set specific dimensions. Now you'll screw the die out two full turns. We do this to remove the crimp, if any, from the setup equation (though just at this particular point). With a 14-turns-in-one-inch thread pitch, or 14 TPI, this action moves the die up approximately 0.144-inch. This means the internal crimp ring of the die is well above where the case neck will be at top of ram stroke. At this point, lock the die lock ring to maintain this position, but don't lock the set screw just yet.

Place a primed and powder-loaded case into the shellholder and set the bullet in the case mouth; here you will see where the benefit of chamfering comes into play. Push the bullet in a bit, though the neck tension should keep you from moving the bullet into the case very far. You just want to set the bullet in a proper attitude so that, when it enters the die to contact the seating plug, it is as straight as possible.

Now, before pulling the handle to raise the ram and seat the bullet, back out the seating plug so you can "soft-seat" the bullet. Soft-seating

Soft-seating the bullet is the first part of the full bullet seating. You can see that the bullet in the left picture, which has been soft-seated, sticks out of the case quite a bit, compared to the fully seated bullet in the right picture. It's important to measure your cartridge overall length after soft-seating and adjust your seating die incrementally until you arrive at the proper dimension.

allows you to get the bullet started, as well as centered and secured by neck tension, so you can then make the needed adjustments to gain the proper "cartridge overall length" (COL). This is another measurement that will be found in your powder loading data manual. Cartridge overall length is important to safe functioning of your reload in your firearm. The COL may vary slightly from bullet type to bullet type for the same cartridge, due to specific functional requirements of the finished round, so, again, do not substitute. Your aim is always to achieve the proper COL for the bullet, case, primer, and powder combination specified in the data manual. This will ensure safety, accuracy, and proper function.

Using the dial or electronic caliper, measure the case from the bullet tip, or "meplat," to the flat of the case head where the primer is located. This will give you the COL measurement of the loaded round after soft-seating. Now you must adjust the seat plug to push the bullet just deep enough into the case to provide the proper COL. The thread pitch of the seat plug is the determining factor in how far you'll screw in the seat plug to achieve this. On my Redding seating dies, the thread pitch is 20 turns per inch, or 20 TPI, which conveniently gives me 0.050-inch per rotation. I will use this as an example.

Let's say my powder data manual specifies 1.264 inches as a maximum COL. My initial soft-seat has left me with a loaded round of 1.361 inches. Knowing the pitch of the thread is 20 TPI and the rotation will yield 0.050-inch of travel, I will start by turning the seat plug two complete rotations. (As there are no marking on the seat plug, I use the number stamped in the top as my reference.) If I was perfect (and no one ever is), I should get a measurement of 1.361 inches. But reality says I overturned slightly and my round would end up at 1.248 inches. Now I must slowly and incrementally turn in the seat plug to reach my goal of 1.264 inches, which means I need to shorten the loaded round by 0.016-inch, which is ever so slightly more than a quarter-turn.

Do not try and hit this number on one try! You will likely over-seat. You're going to have to use the trail-and-error method. I go just under a quarter-turn, check, and then begin a series of very slight bumps of the rotation to get the length I need, measuring each time until I have reached the goal.

Now, if this bullet has a "cannelure," or "crimp ring," and you wish to crimp the bullet into the case, here is where you can begin that process, and it's why you didn't lock the major die lock ring earlier when you were setting up this die. Using your completed, proper-length loaded round, set up the crimp. A "roll crimp" is exactly as the name implies. A raised metal portion of the die's internal design, called a "crimp ring," rolls the mouth of the case into the cannelure or crimp ring embossed on the bullet. Getting the crimp in the right place shouldn't be a problem, as your properly lengthened cartridge should have left you with the cannelure aligned at the case mouth.

To set the crimp, loosen the major die lock ring and screw it up and out of play for the moment. You'll also need to back out the seat plug you worked so hard to adjust and remove it from the operation. Next, raise the ram, with the loaded round in the shellholder, to the uppermost position. While the case is in this position, begin screwing the die down until you feel contact. This is the crimp ring in the die contacting the case mouth. Now, back off the ram and screw in the die another eighth of a turn; with the 14 TPI of the die body, this equates to about 0.010-inch. This should be enough to provide a good crimp. Lock the major die body lock ring and crimp the case. Inspect to see that the crimp has sufficiently rolled the case neck in and, with the completed cartridge

Once you have a finished round, it's a good idea to test it for concentricity. Some reloaders simply roll a round gently across a flat, level surface to see that it rolls in a circle around its point, but for more precision, a concentricity gauge is a handy tool.

in the shellholder and the ram fully up, slowly and carefully screw the seat plug back in until you feel contact on the bullet ogive. Lock the seat plug in position and lock the set screw in the major die lock ring. Now all adjustments have been made to both seat to proper length and crimp in one step.

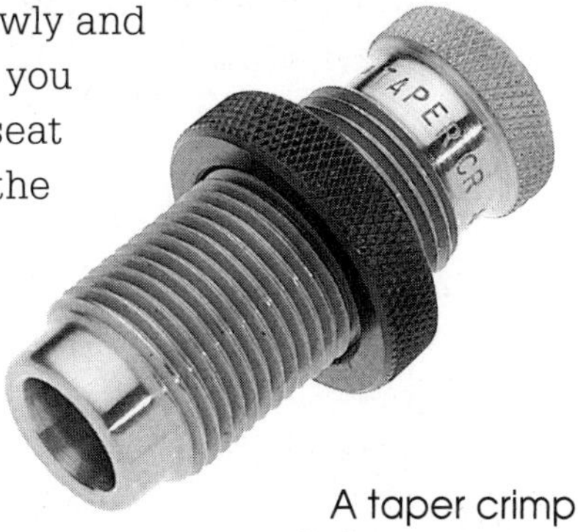

A taper crimp die is used when a cartridge is intended to headspace in the firearm via the case mouth.

I previously discussed the roll crimp and how it's applied to bottleneck cases. In straight-wall cases, though, you can have both roll crimps and taper crimps, and their uses are based on how the individual cartridge case headspaces within the chamber. On most revolver cartridges, the headspace of the gun is dimensioned off of the rim, thus, these cartridges work well with a roll crimp. Many auto-loading pistols, on the other hand, use a rimless case and so headspace off the case mouth surrounding the bullet. In this latter case, a roll crimp wouldn't work, because rolling the case mouth in would not allow the square case mouth to contact against the corresponding dimension in the firearm chamber; this allows the potential for the case to be pushed too far forward, causing malfunction and, possibly, a safety hazard.

For cases where a roll crimp won't do, you'll use a taper crimp die. This is, generally, a distinct process and adds a fourth step to the loading of straight-wall cases. In other words, you'll size, expand, and seat as usual, but it will be necessary to crimp in a completely separate operation. Many accuracy buffs prefer a distinct crimp, as the bullet is being pushed out ever so slightly as the crimp is applied in a one-die scenario.

To set up and use the taper crimp die is very simple. Screw the die partially into the press. Put the loaded cartridge in the shellholder and raise the ram to the topmost position of its stroke. Begin screwing the die in until you have contact with the case. Lower the ram and turn the die in again, this time approximately an eighth of a turn. This will provide about 0.010-inch of crimp. Remove the case and push the bullet against the side of a bench. If the bullet does not move, you are good. If there is some movement, turn the die in slightly again and recrimp until the bullet holds firmly when tested again in the same manner. Once complete, set and lock the major die lock ring and the process is complete.

PART II:

Shotshell Reloading

SECTION D:
Shotshell Reloading Fundamentals

13 What's Shotshell Reloading All About?

Reloading for shotguns is somewhat less complicated than reloading for metallic cartridges. Indeed, some would say it's not much more than dropping in powder, a shot wad, the shot itself, and closing the top of the hull over the insides so they don't fall out. That would be true, to an extent, but that's really far too simple a description of the process—and you know that much already or you wouldn't be reading this.

Successful shotshell reloading requires that you pay attention and

Shotshell presses are most commonly found in a progressive setup, as in the press on the opposite page, but single-stage setups can be had, too.

that you follow the steps to reloading the shotshell in the order they should be followed, but it's not so complicated a process that you'll have to compute exotic equations or decipher arcane hieroglyphics. Building shotshells means following recipes, just as you would when preparing a dish in the kitchen. Sixty minutes at 425 degrees Fahrenheit means exactly that. If you err and turn the oven to 400, your dish will be undercooked. Go to 450 and it will scorch. In the same vein, if you underload the powder, the shot will not achieve its most effective velocity; overload and you can damage your shotgun or, worse, blow out the breech and hurt you or someone standing nearby. This section of the book will explain the ingredients that go into a modern shotshell and how to put them together. All you have to do is follow instructions.

The last couple decades have seen a proliferation in three types of loads that have made shotshell reloading more relevant than ever.

First, there has been an expansion in the variety of shotshell hulls available for reloading. It used to be that the "Big Three" of Winchester, Remington, and Federal were pretty much the only makers that provided a diverse lineup of shotshells and, thus, those were majority of hulls reloaders used. Now you'll regularly find boxes of B&P, Eley, Kent, Fiocchi, and Wolf shotshells at gun clubs and on the shelves of local retailers, and you'll also find loading data for them in abundance.

The second development in shotshells is that manufacturers have recognized the growing interest in lighter (smaller shot payload) loads that still get the job done. There was an era in the '70s and '80s when every shell strained for maximum power and greatest possible impact on the target end. Of course, that meant maximum physical punishment from blast and recoil to the shooter. This is no longer the case. While the big 10-gauge shell and the magnum 3½-inch 12-gauge hold court among waterfowlers and some numbers of predator and turkey hunters, for most other shotgun applications, it is now relatively easy to locate shells loaded with less powder and less shot rather than more—minis versus magnums

Finally, there has been an explosion in the development of non-toxic loads. Offerings of bismuth, tungsten-compounds and, now, *much* improved steel loads have greatly refined the selections in green shooting. Even the humble wad is evolving toward complete biodegradability, a boon for shooting clubs looking to lower cleanup costs, as well as public shooting lands where the standard plastic wad and lead shot aren't viewed as being environmentally friendly.

This all sounds grand, and a lot of it is, but, as diverse as the shotshell market now is, can you find the exact load you want exactly when

Reloading will introduce you to the intricacies of a new science. Dealing with the components that make up a shotgun shell teaches you about burn rates and dram equivalents and skived hulls. There's a lot to learn, and studies indicate that as we get a wee bit older, it takes us longer to wrap our minds around and master new skills and information. So, why wait?

you want it? Retailers tend to stock fare regional in nature—say, a proliferation of pheasant loads in the shops in Pierre, South Dakota—or whatever they got a good deal on when buying in bulk. That doesn't help you if you're looking for a certain sporting clays load or a 20-gauge turkey shell, but that's all the more reason to begin reloading!

Reloading isn't a new concept. Muzzleloaders, the original shotguns, required every shooter to load their own. In those old days, people learned to load from pa and ma and they stuffed their patches, powder, and shot right down the barrel from the muzzle to the chamber. Once breechloading guns became available, that changed. Shooters could purchase ready-made cartridges that would fit their guns. Muzzleloaders and reloading declined rapidly in popularity after that, and while muzzleloading shotguns are today in use by only a small number of shooters, the practice of reloading, thankfully, is seeing a marked resurgence in its popularity. It's that modern popularity and its bevy of advanced components that this half of the book will address.

COST

Many reloaders will tell you they save money by building and rebuilding their own shells. It should be obvious that, considering the cost in components and the time and effort it takes to reload, you cannot save money if you are shooting only a few boxes of shells a year. In fact, if you don't shoot often, you should probably not bother to reload, because you may never recoup the initial investment. Let's look at an example of why this might be.

Nowadays, a MEC 600 Jr. Mark 5 single-stage reloader retails about $200. After that, you can expect to pay about $11 for a bag of 100 Fiocchi 12-gauge 2¾-inch hulls, and 14 ounces of Hodgdon's Clays powder is another $15. A 25-pound bag of No. 7½, 8, or 9 chilled lead shot can run you $17 and up, more—sometimes significantly more, say, $23 to $55 a bag and higher—for high-antimony (harder) and coated or plated shot. A bag of 500 wads will run you $10 to $12 dollars, a box of 1,000 primers goes from $30 to $50, and 500 once-fired AA hulls (if you have to buy hulls to get started), will run you in the $50 range from a supplier like Gamaliel shooting supply. Add a few knickknacks that your local dealer or a reloading buddy recommend, and perhaps the cost of all those factory rounds you shot to accumulate you're stash of once-fired hulls, and you are going to spend several hundred dollars just for your initial setup. I don't know about you, but to me that's an awful lot of cash if all you're going to load is a couple dozen boxes of shells over the course of a year. Compare that to a case (10 boxes) of Winchester AA target loads runs that run about $90 on Cabela's website, and you won't even have to worry about shipping if you have one of their stores near enough that you can stop by and pick it up. Even at three cases, you're still spending less than you would getting a shotshell reloading bench properly setup.

The avid shooter, on the other hand, the guys and ladies who shoot skeet league every Wednesday weeknight and sporting clays tournaments every weekend, the duck and goose hunters who are hunting across multiple states and multiple limits, and the upland hunters who are really putting the miles on both their trucks and their bird dogs, now those, those are the shotgunners who can save some cash by reloading.

Take a dedicated tournament shooter, for instance, and do the math with that same case of factory new AA target loads. Let's say you shoot two 50-round sporting clays courses or four rounds of skeet for practice during the week and a 200-round tournament every other weekend. That's more than a case a month! For shooters spending that kind of time behind the gun, reloading can absolutely make sense, even more so when groups of shooters can get together and buy cases of wads or a pallet of shot and really cut down the overall cost of components. Add to this fact that the more you shoot, the more quickly you'll amortize the initial cost of your press setup (even though many reloaders find they'll eventually buy more complex and expensive gear as their shooting progresses and they experiment with more diverse loads).

Practically speaking, the people who shoot often enough to get the best value out of their reloading gear are usually target shooters, those trap, skeet, and sporting clays enthusiasts. Dedicated waterfowlers aren't far behind; though their annual number of rounds won't approach that of a dedicated skeet competitor, the waterfowler's non-toxic loads are inherently more expensive and, coupled with extended seasons and limits for geese in many regions, that can turn into a lot of money going out the door. Again, reloading can cut that cost down. Regardless your

Target shooters have a reputation for shooting many more rounds than hunters. Over a good season of deer hunting you may shoot just two or three rounds, and a dove hunter would run through maybe two boxes of shells for his limit. A sporting clays or trap shooter, on the other hand, will burn up five boxes of shells sometimes a couple times a week and several cases a month!

If you shoot more than a few boxes of shells a year, you must explore this fascinating hobby. Scott Richardson, an avid shooter from Gainesville, Florida, built a temperature- and humidity-controlled room for his many reloading presses. That's some dedication!

sport, I think that the people who enjoy shooting most and who stick with it the longest, making it a lifetime sport, are those who have a diverse range of interests. They are clay shooters and they are hunters, too. Reloading is ideal for these committed, multi-tasking gunners.

There's another group of shooters who should consider reloading, if they're shooting any kind of volume. While the 12-gauge shotgun is the standard in North America and, perhaps, around the world, there are still plenty of 10-gauges and millions of sub-gauge 16-, 20-, 28-gauge, and .410-bore guns in the hands of shooters everywhere. Folks who shoot sub-gauges because they enjoy the challenge, prefer the softer recoil or, perhaps, because they're training a spouse or youngster, will almost always find that the shell costs for those gauges are higher than for their 12-gauge. For instance, a 25-round box of Remington's STS target shotshells in 2¾-inch No. 8 from Bass Pro Shops retail about $9.50, while the comparable 28-gauge load runs almost 50-percent more at $13.99. That's a *huge* difference in cost for the same 25 shots! Again, you might shrug your shoulders and say, "So what?" if you only use that 28-gauge a few times a year for grouse or woodcock, but a clays competitor committed to shooting sub-gauge events will surely see the monetary benefits of reloading the smaller gauge.

Now, I know some of your are thinking that bigger shells containing more powder and more shot would cost more, but this is not the case. Why? The laws of supply and demand tell us that because many more 12-gauge shells are purchased than sub-gauge shells, manufacturers gear up their primary production lines for the 12-gauge. Although sub-gauge shells are not necessarily an afterthought—they are still commercially important—they will be secondary on the assembly line and produced and sold in smaller volume, hence, sub-gauge shells cost more.

Another angle to consider is that savings don't always have to be about money. There's your time and effort to consider, too. Consider going to the store for a $10 box of Federal No. 4 turkey loads. Depending on your home press setup and reloading accessories, you may be able to punch out half-a-dozen turkey shells and then re-set the shot charge bar for your No. 7½ or 8 league trap loads in 10 minutes. That's convenience!

CONTROL, CONVENIENCE, AND FLEXIBILITY

More than the potential cost savings, I believe that a much better reason to reload is that manufacturing your own gives you greater

control over your shooting. If you are a careful reloader—and you should not approach a reloading press in any other manner—you can produce loads that are completely consistent from one shell to another. Consistency is important to "groove" your shooting, to give you the confidence to know that, when you pull the trigger, you know exactly where your shot will go (how it will pattern), and how quickly it will arrive where you have pointed it.

Once you identify your needs and shooting interests, you can easily program your reloading gear to produce the shells you need, because reloading also gives you versatility. It is quick and easy to make changes and build, for instance, a dozen different shells with which to experiment. For instance, if you're a sporting clays shooter and find it irritating to change screw-in chokes between stations, you can load up different batches of No. 7½, 8, and 9 at different speeds to address the wide variety of targets and distances you'll encounter. The reloading experience also allows you to experiment with different components. If you have never shot some of the spreader wads, for example, it should be easy to buy a small batch from an Internet site, a friend, or your local reloading supply dealer and load a half-dozen shells, then see how they pattern with an eye to those fall days of woodcock hunting. Indeed, once you start thinking about the possibilities after you begin reloading, it's not hard to see how your shooting opportunities might be expanded.

Having a reloading station at home can help you overcome the recoil dilemma, too. Recoil is a serious issue among all levels of shotgunners from professionals to novices. All shooters, regardless their firearm, are sensitive to and are eventually affected by recoil. It is only a matter of time before you must deal with it, because recoil causes flinching (sometimes called "target panic"), and flinching causes you to miss what you are shooting at, whether it's a clay disc or a grouse thundering out of the quakies.

Reloading can help tame recoil and muzzle blast by letting you find shell recipes that will accomplish your objectives with lighter or different loads. Many shotgunners discover that lighter loads accomplish the same killing results on birds or clays as their heavier standard loads do, but with the benefit that lighter loads punish their bodies less. As a build-your-own specialist, you can create literally dozens of loads, experimenting with different primers, powders, shot sizes, and wad types, as well as different shot charge weights (anywhere from powerful 1¾-ounce loads down to the relatively small 7/8-ounce charge in a standard 2¾-inch 12-gauge load). Plus, reloading gives you the opportunity to experiment with a few shells at a time, rather than buying a whole box of 25 that you may never use beyond the first five shells—with reloading, essentially, you can try before you buy!

OTHER NICETIES

For some practitioners, reloading is just a means to an end, merely the simple act of creating shootable shells. For others, reloading and patterning become a quiet passion. It deepens their interest in all facets of shotgunning, involves them in a wider community of shared interest, promotes an understanding of and ability to negotiate an interesting and unusual technical field, and becomes a path of deepening commitment to shooting. Hunters find themselves shooting a little trap and skeet, and sporting clays enthusiasts begin tinkering with and patterning turkey loads. Whatever your sport, whatever new pursuits you discover, reloading for your shotgun can help you achieve success.

14 Shotshell Components

All shotshells are essentially the same and all are very different, in the same way that all cars are the same, but none are identical. Shells have the same basic elements and those elements have the same basic functions. Nevertheless, there are practically endless combinations of these elements, and reloading just one of every possible combination would be a lifetime of work.

Besides the tools needed to put them together, you need five essential ingredients to build a shotshell. Those five components are:

- A hull or case, which acts as a container for the other components.
- A primer. This is held in the base of the hull. The gun's hammer strikes the primer when the gun is triggered, and the primer explodes, igniting the powder above it.
- A small volume of powder that burns rapidly and becomes a large volume of hot gas to push the shot through the barrel.
- A wad and/or shotcup to hold the shotshell's pellets and protect them from the powder burn.
- Shot to fill the cup and shell.

Just for your library of knowledge, and though these things are beyond the scope of this book, as they involve more advanced reloading practices, know that some specialty loads may have additional internal spacers to help fill the hull and balance the load, and there are specialty loads that contain a buffering agent that helps protect the shot from deformation.

THE SHOTSHELL HULL

The hull's duty seems obvious, but, in fact, it has quite a few functions and all of them must perform precisely as designed to deliver quality shooting.

The primary function of the hull is to package the powder and shot in a neat and complete unit. Since relatively few shotgunners are black-powder enthusiasts, most of us just want to shove a shell inside the chamber or in the magazine and pull the trigger. When we are in the field or at the range, we want to pull the trigger—a lot—and we do not want to assume that the shell is anything other than okay. That's why we use modern shotguns instead of muzzleloaders now, so we don't have to worry about rain getting powders and primers wet, or spilling powder and shot while we're trying to reload. Today's shell provides convenience, safety and, with a few exceptions, the capability of being recycleable.

As similar as they are in appearance, all hulls are not alike. For instance, in the discount marts, you will find numerous boxes of low-priced shells, often called "promotional loads." These lead shot-filled shells are excellent for one-time use, but their hull should not be considered seriously for reloading; you might, *might*, get one or two reloads out of them, but they're really not worth the effort. Why? Well, there's a reason these shells are cheap, but one of the primary problems associated with reloading these hulls is that they are usually constructed with paper inserts inside at the base of the hull. These inserts quickly detach with use beyond their original, factory loaded firing, and that detachment can lodge in the gun barrel—and *anything* lodged inside your barrel is going to be a problem.

HULL BODIES

A little bit of history here. Hulls have evolved significantly during the last century. Originally, the self-contained shotshell hull was brass from top to bottom. All-brass shells were cumbersome and expensive, though. Paper quickly replaced the all-brass construction, though a rigid brass base was retained to help seat the base wad, hold the primer, contain the shaped paper hull, and provide a surface that allowed solid grip by the gun's extractors after firing.

During the 1960s, a remarkable development in shotshells took place, with the introduction of plastic cases. Today, almost all shotshell cases are plastic, which is much easier and less expensive to form into a shotshell case than is paper. Plastic hulls offer greater water resistance than paper, but more important for reloaders, plastic maintains a superior "crimp memory," eliminating the fraying associated with the edges of layered paper hulls. Plastic used for hulls is more resistant to heat than paper hulls, too, and is stronger for the amount of material required. Today, and to my knowledge, Federal Premium Ammunition is the last

Shotshells used to be comprised entirely of brass. They were sturdy enough, but expensive and physically heavy. Paper replaced the case first, and plastic quickly replaced paper. Though some specialty suppliers still make all-brass shotshells available (mostly to vintage gun shooters), mostly all we're left with is the metal head on a modern shotshell as a reminder of how far we've come.

remaining American ammunition manufacturer producing paper-hulled shotshells, those in its famous Gold Medal competition line. These hulls are reloadable; if you haven't shot through enough factory-fresh Federal Gold Medal paper hulled shells to give yourself a good supply of empty hulls to work with, you will likely find batches of once-fired Gold Medal hulls for sale, though be prepared to find them priced higher than their plastic counterparts. Ballistics Products, Inc., for instance, sells 100-count bags of once-fired, unprimed Gold Medal papers for about $12, and primed Cheddite paper hulls (Cheddite is a French producer of primers and hulls) go for nearly $16 a bag (the primers do up the price a bit, and you will incur a hazardous shipping charge, something else to consider, when you're totally up your costs). In comparison, unprimed 100-count bags of plastic Remington STS hulls go for about $10 and unprimed Rio hulls at a 100-count were on sale for $4 when I typed this up (in early 2014).

As good as plastic is, there is considerable pressure from outside the industry to shoot biodegradable components. (In my hometown in Florida, a city known for its liberal politics, the new Gainesville Target Range requires shotgunners pick up not only their hulls, but their wads,

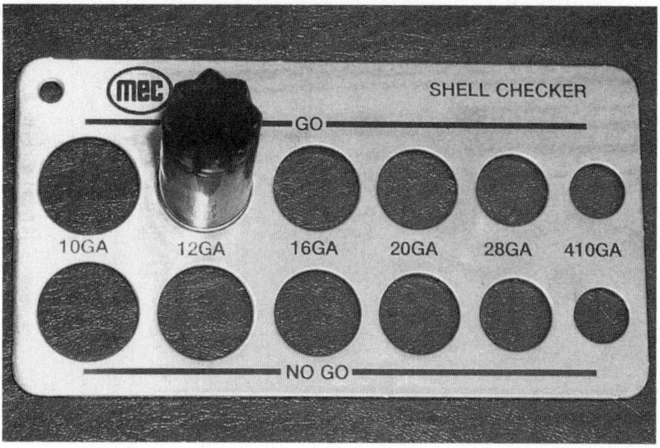

You can easily measure the precision of hulls and reloaded shells with a shotshell checker. Precision machined holes, labeled GO and NO GO in this stainless gauge-checker plate from MEC, test for accurate shell size and roundness.

as well.) Manufacturers are developing more eco-friendly products and the good news is you will increasingly find them available as reloading components. Kent/Gamebore, for instance, developed a photodegradable wad. It's still plastic, but it incorporates properties that cause accelerated breakdown. Along the same lines, the company's Gamebore line has a 2¾-inch, biodegradable, varnished paper shell, and the load also contains a fiber shot cup. These are just a few examples; a discussion of non-toxic shot is a subject nearly for its own book. This movement may, sooner or later, affect handloading in a giant way. For the present, just be aware that such components are out there and that they may require different handling for reloading than do plastic hulls. Beyond that, plastic is still the dominant hull material, so that's what the discussion in this book will be contained to.

The height of the brass base on a shotshell is not an indicator of its contents or power. A high-brass base shell does not necessarily contain a heavier load or greater charge of powder than does a low-brass base shell.

HULL BRASS

Looking next at the base of the hull, you'll see a brass cup that stabilizes the hull and other components. Curiously, this "brass" cup is often not brass at all, but rather a lightweight steel alloy colored to look like brass. Why? Shotgunners just like it

that way. A few years ago, the now-defunct shotshell manufacturer Activ tried aggressively to market a no-brass, all plastic case. This case was entirely functional, and it was suitable for reloading, as well; it incorporated a small metal ring molded into the base to grip the primer. Nevertheless, sales results for the unusual shells weren't pretty.

When I bought my first shotgun, about 25 years ago, I learned that there were two kinds of shells, high base and low base. This referred to the height of the brass base. It was generally understood that a low base equated with low power and a high base with high power. Consequently, my buddies and I purchased high base shells for pheasants and waterfowl, and low base rounds for grouse and woodcock. What a surprise it was to learn there is no essential relationship between the height of the base and the power of the load! But myths die hard, so, today, we still have high-brass magnum loads and low brass dove loads.

The metal visible up the side of the hull is designed to stiffen the shell, to give the extractors a firm shelf on which to grip and, especially with paper hulls, to provide a firm base of support for the load's components. During the first half of the twentieth century, the size of the brass on a shell varied as manufacturers experimented with new paper and plastic hulls, new base wad materials and heights, progressive powders, and various configurations for consolidating the elements of particular loads. (One of the difficulties with new hull materials was in finding component combinations that would best contain the gas pressure from the burning powder without leakage around the seal or the base wad.) Today's hull makers vary the height of the brass for the same reason they use differently colored hulls—different sizes help them and their customers distinguish between different types and sizes of shells in their lines. As a final note, a sturdier hull base, one wrapped in metal, albeit lightweight and relatively soft as metals go, may have advantages when shot through gas guns and pump-actions, which tend to extract shells with greater force than over/unders or side-by-sides. For gas and pump guns, you need a shell rim to be made of a reasonably strong material and firmly attached.

Yup, these are high-brass target loads. Let go of the myth--just because the brass is high doesn't mean you've got a pocket full of heavy game loads.

STRAIGHT-WALL VS. INJECTION MOLDED

There are two fundamental designs of plastic hulls in use today, a two-piece straight-sided hull and a one-piece, injection-molded, compression-formed hull.

The straight hull has a removable base wad of paper or plastic that separates the powder from the metal of the base. This shell is typically the thinner of the two designs and is frequently the hull of choice for building powerful hunting loads, where every spare millimeter of internal space is packed with powder and shot.

At the base of a straight-sided hull is the interior base wad. It is somewhat rare that this is a loose paper wad, as it was in days past. Subjected to the intense flash of heat when the primer explodes and ignites the powder, a paper base wad can easily detach and fall out, deteriorate, or blow forward and stick inside the barrel. If you choose to reload straight-wall hulls, you must inspect the base, whether it's fiber or plastic, for adhesion of the base wad and discard those hulls that aren't intact. Today, many base wads are plastic and look like washers. Their functions are to separate the powder from the base, to elevate the powder to the top of the primer, and to help seal the bottom of the shell. You do not want a gas leak anywhere, but, if one does occur, it can be especially disturbing around the base of the hull.

One thing to note as you experiment with straight-wall hulls is that you'll see plastic base wads of varying thicknesses. This allowed the original manufacturers to offer a huge variety of load combinations without mandating extreme component adjustments, i.e., they simply changed base wads to change the height of the powder column, rather than switching out ever so many different powders.

Crafted differently than the straight-wall hull, the injection-molded or compression-formed hull uses an integral base wad; the hull wall itself tapers to the bottom toward the primer hole, so that the thickening curvature of the hull is what separates the powder from the base metal. Winchester's original AA hulls were a good example of the more common compression-formed tapered hull. (Winchester changed the design of this hull a few years back to a two-piece hull, but the company states the two are completely interchangeable regarding reloading data and practices.) Typically, the thicker, injection-molded shells are the choice of hull for competition loads. Reloaders like them because of their longevity, i.e., they can be reloaded several times, sometimes dozens of times, before they have to be discarded.

HULL CRIMP

At the opposite end of the hull's base is the crimp, and no matter whether the hull is paper or plastic it needs to be positively crimped. One purpose of the crimp is to seal the end of the shell to prevent the shot from falling out and keep dirt from entering. The crimp also keeps the powder and shot properly packed for that micro-second when the primer ignites the powder and pressure begins to build.

The crimp is a patterned fold. Its design is essential for proper powder ignition and controlling the burn rate. Varying the depth of the crimp or otherwise changing a pre-established fold when you reload can quickly and surprisingly affect your shell's pressure, so mind your crimp as you do the rest of your components and stick to the recommendations of the recipes you use.

A few years ago, two types of crimps were common, the roll crimp and the star fold. The roll crimp dates from blackpowder days. Blackpowder was bulky, at least compared to today's smokeless powders, so it needed all the room it could be afforded inside a shell. Everything was packed in tightly and a small over-shot card (also called a wad) topped off a roll-crimped load before the crimp was applied. The crimp, when applied, rolled the hull firmly back on itself and down to the card, thus holding the powder and shot firmly in place.

Beautiful, six-fold factory crimps on Federal No. 6 12-gauge and 20-gauge shells (left and top), and a Wolf No. 8 12-gauge. You can build crimps just as good looking and, indeed, to be effective you must make this part of the reloading exercise a priority.

With the advent of more efficient smokeless powders, less hull length was needed to contain the powder, because less powder volume in a smokeless loading could accomplish the same or better results than did blackpowder. As a result, more hull was available for sealing the shell. The over-shot card was dispensed with and the final quarter-inch of standardized paper or plastic shell was simply folded over toward the middle. Today's final crimp depth is about $^{1}/_{16}$-inch and has either a six- or eight-segment fold. Your shotshell press, which we will be discussing in another chapter, should accommodate crimp starts of either configuration, and you should use the right one depending on the number of folds your hull originally had.

Is there a difference between the six- and eight-segment folds? Except for the number of leaves or folds, no, but it is believed that the eight-segment fold holds a little tighter and is, therefore, a little better for the smaller shot sizes of No. 7½, 8, 8½, and 9 used in target and small-game loads. There is a tendency for the six-fold to be used with larger shot in hunting and field loads. Also, the smaller shells in the 28-gauge and the .410-bore use a six-fold crimp, and though it seems counter-intuitive, the large shells of a 10-gauge also use a six-fold crimp. Many experienced reloaders recommend that, when you work with a new (not previously crimped) hull, consider using a six-point fold starter rather than an eight-point, if you have a choice. The six-fold is easier to work into a fresh hull and usually realigns more easily.

HULL INSPECTION

Because hulls are so important to successful shooting, you must inspect them carefully before you begin reloading. Initial sorting separates gauges and lengths. After that, you can sort for brand and type. Just because you are holding a 12-gauge hull doesn't mean you can load it with just any 12-gauge recipe. Seemingly small but ballistically significant differences in hull material, design, capacity, and wear will affect loads profoundly. Winchester, for example, used to advertise that its AA hull would "become the new benchmark of reloadability." In the same ad, the Illinois company suggested that you can achieve at least 15 reloads from a single shell. Let me tell you, while Winchester AA hulls have an excellent reputation among reloaders,15 is a *lot* of reloading. Not only is the hull subjected to violent extremes of heat and pressure every time it's fired, but also the crimp must fold and hold precisely each and every time the shell is reloaded.

One of the biggest issues you'll likely face in shotshell reloading is forcing yourself to discard hulls that may not be in the best shape. But do it. Shoot a box or two of new factory shells or buy a bag of new hulls for every couple hundred reloads you shoot and ensure you have a regular supply of viable hulls to replace the ones you'll eventually need to discard.

What I'm getting at here is that nothing is going to work better for you, either in your reloading press or in your shotgun, than will perfect shells. Many writers recommend you begin reloading with a bag of new or once-fired and pre-sorted hulls that can be purchased from your local dealer or via a known Internet vendor like Cabela's. Still, it's well recognized that most shotshell reloaders use hulls well past their prime and, so, wind up trading performance for economy. No one would consciously choose to make this trade, but we reload in part to make our money and components last longer than one shot; there is always that push-pull dilemma of getting "just one more shot" from a hull, but

Compare a few spent wads to new wads (top and bottom right) and you will quickly realize two things: A). why they are not reusable; and B). the power of the setback inside the hull. Shot leaves permanent indentations in the base and sides (petals) of the plastic wads, clearly visibile in the photos above.

worn-out shells destroy shot-to-shot consistency. Hulls with compromised structures will leak gas and cause the loss of both pressure and velocity. Worn-out crimps will leak the shot beneath them (every target shooter I know has gotten a small pocketful of shot in their shooting vest from such loads, at one time or another). Fortunately, hulls occasionally give you easily discoverable clues about their readiness for retirement.

Hulls begin to fatigue with their first loading. Repeated reloading and firing eventually causes the seal between the base of the paper or plastic hull and its brass base, or the seal with the base wad, to deteriorate. In some hull types, the process of base degradation can be rapid and extreme, and therefore more noticeable than in others. Other symptoms of hull fatigue include plastic walls becoming brittle near the top or at the crimp, ultimately developing hairline cracks there that leak gas. You must toss such hulls immediately.

THE PRIMER

This 12-gauge base clearly shows the spent, dented primer that will have to be removed and substituted with a fresh primer during the reloading process.

Struck by your gun's firing pin and instantly super-heated, the tiny amount of chemical inside the primer in the base of a shell explodes. This supplies sudden, intense heat to the propellant (gunpowder) by driving tiny white-hot particles up and into it. The burning particles launch the propellant on the brief but glorious arc of its burn.

For maximum efficiency and effectiveness, a primer must offer the precise heat that a particular propellant needs. Bulky, slow-burning powders, for instance, require a specific type of flame in order to ignite and burn properly. Too much heat and flame unnecessarily raise the early chamber pressure, while insufficient heat and flame do not ignite a large enough portion of the propellant for pressure to rise sufficiently before the chamber pressure is then decompressed by the load's movement and its decreasing confinement.

The story of the primer began about 1807, when Scottish hunter and inventor A.J. Forsyth discovered that a particular mixture of chemicals produced an explosion when struck. He realized that, if the reaction was contained and channeled, it could be used to ignite powder charges, and he eventually used his knowledge to create what was called a "pill lock" ignition system. His was the first step along the way to today's modern primers, a path that wound through England and America for more than a hundred years.

Early primer components were effective, but also terribly corrosive. Fulminate of mercury caused brass cases to become brittle. Potassium chlorate left thick deposits like common table salt inside a firearm, making bore cleaning necessary within hours of shooting. Today, the No. 209 primer is the standard for shotshells and, thankfully, a decided improvement over those early tries.

Like powders, individual primer brands have characters all their own. Some burn longer than others, while others burn with greater in-

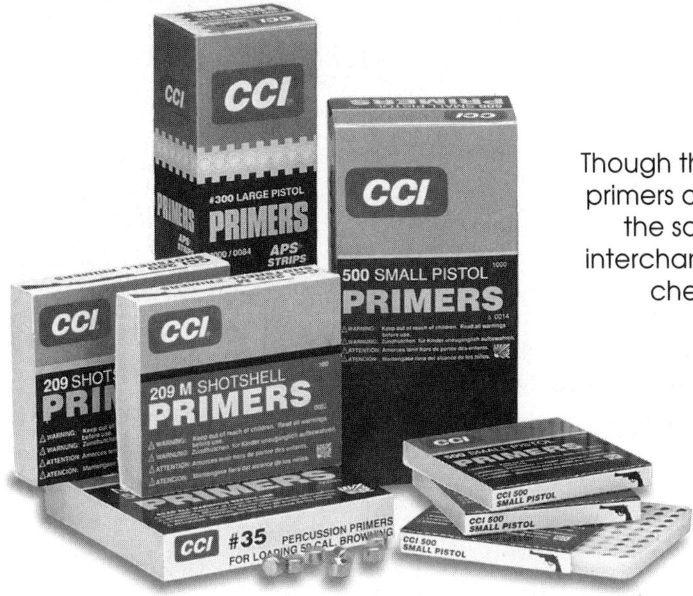

Though they mostly look alike, primers across brands are not the same. They cannot be interchanged without double checking load formulas.

tensity. Some have a longer spark and produce heat over a much longer time, something known as the "flame's duration." In general, target loads do not need much spark, because the propellants in them are in the fast-burning, easily ignited category. Hunting loads, however, may require a great deal of primer boost and heat to get their slow-burning propellants cooking. On a cold day, the need for tight crimps and warm primers can be critical, to avoid sputtering ignition; for these loads, Magnum primers were developed and are especially hot. (Generally, however, magnum shotshells derive their power not from hotter primers, but, rather, heavier shot payloads.)

What this all boils down to is that, as it is with hulls and powders, *never* substitute one brand of primer for another. If the recipe says use a Remington primer, that's what you use, period. It is relatively easy to crank up the pressure in a load to well over the acceptable and safe prescribed pressure with only a primer swap. Tests indicate that some common target loads, for instance, can change by as much as 3,500 psi with only a change of primer. Such a change stresses your gun unnecessarily and potentially enough to damage both it and you, and you're also going to feel the effects of the higher pressure when it comes to the recoil of such an over-pressured load.

THE WAD/SHOT CUP

In the reloading process, which will be covered in much more detail in a later chapter, you start with a hull, punch out the old primer, insert

Just as it is with primers, powders, and hulls, all wads are not the same. Certainly, they vary by gauge, but they also vary by load capacity. A Winchester AA wad is different than a Windjammer, which is different than a BP STS. Other than those aftermarket wads (like those from Claybuster) that say a size is a direct substitute for a specific factory wad, make no substitutions with this component.

a new one, and then add the powder (I'll be covering powders in another chapter). Now it's time to insert the wad.

A wad is a necessary item in a shotshell. It's also a curious one. On the one hand, it is something very simple, but it is a component absolutely crucial to good, consistent performance.

As you might anticipate by now, one wad is not like another. In fact, during the past 150 years, the wad has evolved as much as any other component of a shotshell. One way it has evolved is from being comprised of paper or thick circlets of felt and cardboard to their current, most common composition of plastic. Beyond that, today it's experiencing a further progression toward employing biodegradable materials such as seen in those from Kent/Gamebore that I mentioned earlier. Another way that it has evolved is from a flat disk or several stacked disks to, well, what is really a pair of opposite facing cups with a springy cushion connecting them.

Let's talk about wads in more detail. The wad has two primary functions. It seals the powder away from the shot and prevents the burning gases from leaking through the balls of shot or around the sides, something that would otherwise diminish the unitary force of the shove to the shot payload down the barrel that those gases provide. In this sealing, your wad must fit smoothly and with great precision against the sides of the hull. It cannot be so tight that it causes undue pressures from the burning powder to develop (i.e., it cannot act more like a stoppage), but also not so loose that it does not completely seal off the gas.

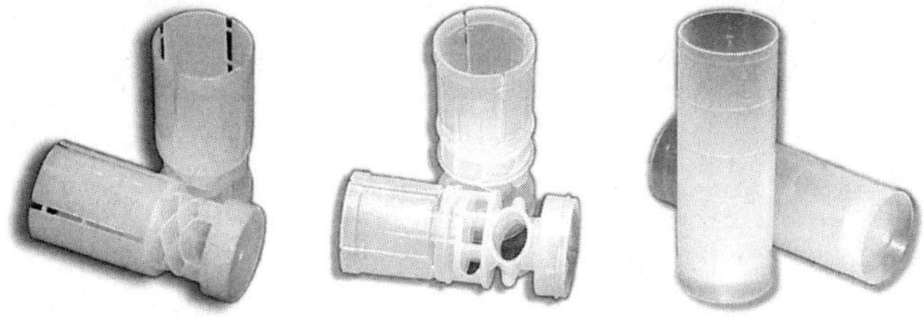

These are just three of the dozens and dozens and dozens of wads available for your reloading. They take on all different shapes and capacities, though the most common types you'll see usually have some sort of "spring" between the shot cup and the over-powder cup, such as can be seen in the left and middle examples.

Because the cardboard and felt wad circlets of a hundred years ago have evolved into a much more impressive and functional shot cup, the wad itself has accrued additional functions. By cupping the shot, it prevents significant contact between the accelerating pellets and the smooth steel of the barrel. This may not be such a big deal with lead shot, but steel shot and today's very hard non-toxic alternatives have been known to scratch older barrels made from soft, relatively thin steel, as well as damage barrels with fixed chokes, especially tight chokes.

By minimizing the contact between your barrel and the pellets in a load and the friction that would result from such contact, the shot cup also helps to keep shot pellets from deforming. Lead pellets are soft and, when they make contact with the hard steel of the barrel, they can flatten in spots. Such deformed pellets are destroyers of good patterns, the so-called "flyers" you'll see on a patterning board.

Once the shot cup has exited the barrel, it falls away quickly. Most modern wads are designed with petals that opens like a flower, the sides folding quickly backwards after the wad exits the barrel. Air resistance causes the shot cup, its petals now open like a parachute, to fall behind the pellets almost instantly, and the wad usually falls to the ground within 10 to 15 yards of the shooter. (Older and experimental shot cups that lacked these petals often fell to the ground with some pellets remaining inside—not a good way to get your pattern to the target!) Though a topic for advanced reloading practices, un-slit wads (those without pre-cut petals) are available, the intent being that the reloader will slit them to a particular load and pattern preference.

In separating the burning powder from the lead shot, the wad or shot cup also helps prevent the extreme heat of the burning gas from melting the lead. Lead melts at 621.5 degrees Fahrenheit and boils at

3,164 degrees. According to Mike Daly at Hodgdon, today's smokeless powders generate heat in the range of 2,800 to 3,200 degrees Kelvin. That computes to a range of 4,575 to 5,300 degrees Fahrenheit. That is easily hot enough to melt your lead shot, if the temperature is sustained for more than a few seconds. Of course, your lead shot is by no means exposed to that heat for that kind of time, but, for the time the gas and shot are in the barrel, there's enough heat to fuse some of the pellets if the hot gases slip around the wad and mingle. Imagine the havoc this will wreak with your carefully constructed shot pattern. Disastrous!

Yet another function of the wad/shot cup is to provide a flat, regular surface against which the expanding gases can press outward. A load of shot, even the smallest target No. 9, is "porous," in that there's space anywhere the spheres of shot don't contact one another. Such an arrangement would allow gas to blow through the load unevenly, if the shot payload wasn't backed by a wad or shot cup. So, the wad provides a ceiling against which the hot, expanding gas can press evenly, thrusting the shot down the barrel uniformly. At the same time, the wad seals the gases behind it in an area of limited oxygen, an element, of course, required for combustion. Of course, that limited oxygen supply constrains the propellant's burn rate, and only by forcing the shot and wad down and out the barrel does complete combustion take place.

There are numerous shot cup designs from several internationally recognized manufacturers. When building a specific load, use the specific components called for in the formula. Why? For one, there are several shell types for any gauge. Second, and even though it might not be immediately obvious, some shells are straight-walled while others have tapered walls. A wad designed for use in a straight-walled Federal hull might not work to seal the gases properly in Winchester's compression-formed AA hulls that have tapered walls. Winchester and a few other companies make wads with a smaller, angular base to fit these special hulls.

Using the wrong wad can produce sub-par performance if it results in "powder migration" past the seal, or if significant "blow-by" (gas escaping around a load, rather than pushing with all its force behind it), reduces load pressure and velocity. Also, most modern wads are designed to "obturate." This means that the bell-shaped gas seal of the plastic wad flares out slightly under pressure to create a consistent and self-sustaining seal that will provide your load maximum drive and consistent ballistic performance. This consistent seal is especially important with high-velocity loads, but the problem of inadequate gas seal grows exponentially when shot cups from smaller, tapered shells are forced to carry maximum loads through today's over-bored barrels. Thankfully, this mismatch can easily correct it at the reloading bench.

Today's wad is, essentially, a slick, three-part element, the shot cup at the top end, the over-powder cup on the bottom end, and a center section. This center section is a cushion of sorts, a kind of collapsible spring. Its job is to work like a shock absorber, although, admittedly, an extremely light one, yet a shock absorber that will progressively collapse evenly, uniformly, and without tipping, which would apply greater pressure to one side of the load.

The modern wad is made of a relatively soft plastic. Pick up a couple after you shoot. You will note the permanent impressions of individual pellets on the worn petals, which testifies to the force of the explosive burn the wad must accommodate. It is also caused by the unevenness of the load's movement through the barrel. The rear shot pellets move prior to those at the head of the shell—they get the push of the expanding gases first—and this force shows in the spent shotcup. Writers Kurt Fackler and M.L. McPherson likened the effects of the powder burn to a shock wave, one that caused the bottom layers of pellets to "squash themselves against the upper layers." This shock, they said, was the equivalent of wrapping a towel around a hammer; the instant of most extreme shock and acceleration, an instant that moves the pellets from zero fps to maybe 1,200 fps, is cushioned.

Heavy loads tend to diminish the effectiveness of the wad's cushion. Heavier loads of shot and heavier loads of powder compress the wad much more significantly than do lighter field and target loads. Thus, the energy wave from the powder's explosion is transferred more directly and harshly to the pellets in a heavier load. Probably, these heavy loads need additional cushioning in the way of felt spacers and specially cushioned wads that must be placed immediately below the charge of shot, something you'll need to consider and experiment with if you'll be loading such rounds.

With all this knowledge about wad behavior in hand now, there's something to consider when reloading. When reloading, you want the center section of the wad to compress, but not collapse. A total collapse would ruin that center section's value and could, perhaps, damage or torque the gas seal that sits on top of the powder. So, under the steady pressure of the wad ram—the part/action on your press that pushes the wad down into the shell to seat over the powder charge—the cushion section should predictably compress to accommodate various powder charge heights.

SHOT

Shot choice is a personal decision, but one rooted in the purpose of your shooting. Size No. 7½ rules trap shooting, The Nos. 8 or 9 dominates the game of skeet, and sporting clays competitors use all three and sometimes No. 8½. Small upland birds like grouse and woodcock are taken with shot sizes No. 7½ to 5, ducks and turkeys in Nos. 6, 5,and 4, and big geese in 2, 1, T, and F shot. Then there are the buckshot loads—single-ought (0), double-ought (00), triple-ought (000) and even quadruple-ought (0000)—intended for big-game mammals and defensive uses, and a variety of slugs, which are single projectiles also suitable for big-game hunting and some self-defense applications. Slugs aside, when it comes to shot, you can reload in lead or steel, and there are now many non-toxic shot options in addition to steel available for reloading, should your shooting require them. Just remember that, whatever your choice, you won't know how it's performing out of your gun until you put it on the patterning board.

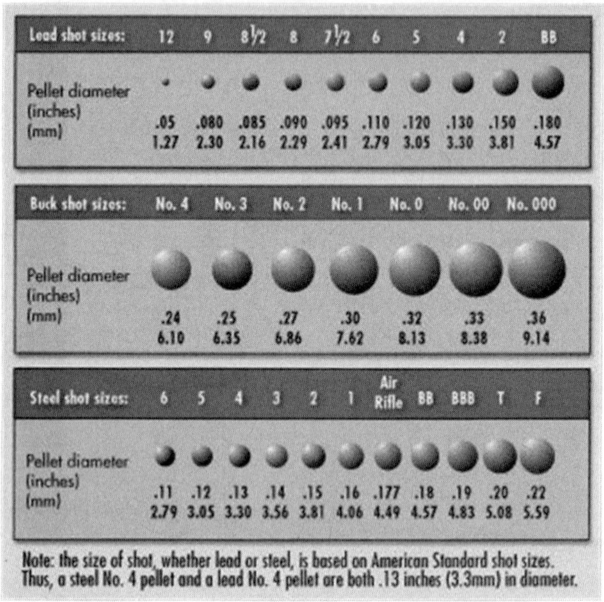

This chart shows the many shot sizes available today and their respective diameters.

15 Shotshell Propellants

To blow the shotshell's pellets out of the muzzle with the velocity required to kill game or break flying targets, your shotgun load must generate a great deal of pressure in the barrel. Achieving this pressure requires that an expanding volume of very hot gas be temporarily confined within the chamber of your shotgun.

As a reloader, you will develop loads and select load ingredients based on the confining volume of—or you could say the compression that develops inside—a particular gun's chamber, a 10-gauge chamber being larger than a 12-gauge, for example, and a 3½-inch shell being larger than a 2¾-inch. Without precision containment and the development of sudden, overwhelming pressure when the hammer and firing pin hit the primer, there can be an explosion, but no shooting. Within milliseconds—millionths of a second—your 00 or No. 8 shot (plus the wad) go from zero fps to 1,200 fps or thereabouts inside the tightly confined space of your shotgun's barrel.

For this kind of shot velocity and the pressure required to achieve it you need what is called a "controlled burn" of a high-energy propellant, not an instantaneous explosion. "Controlled burn" fairly well describes what happens when you pull the trigger: The hammer strikes the primer, which explodes and showers the compact mass of powder with hot, burning sparks. Once it ignites, the powder undergoes a chemical change, transforming from a solid to a burning, expanding gas. As pressure increases in the chamber, the powder continues to burn at an increasing rate, using all the available oxygen as it does. (Curiously, oxygen from the air isn't necessary for the combustion of smokeless powders, since they contain a sufficient built-in oxygen supply with which to burn completely, even in an enclosed space such as the chamber of a firearm.) The hot gas intensifies the compression in the chamber and bore and the cycle accelerates to its climax. Of course, something

These are two of Hodgdon's most popular 12-gauge powders, Clays and Titewad, especially when it comes to target and light field reloads. Clays and Titewad are also rated for competition pistol reloads in the .45 ACP and .38 Special. Clays, Hodgdon says, produces "soft, smooth recoil" and burns "ultra-clean," with a mild muzzle report and excellent patterns. Titewad is a "flattened spherical" powder with low charge weight, "mild muzzle report, minimum recoil, and reduced residue for optimum ballistic performance."

has to give way to the expanding gas. The weakest links in this case are the wad and shot column, which are held at the mouth of the barrel solely by the folded or rolled crimp on the end of the hull.

PRESSURE

One hears about several types of pressure, but, for handloading, there is only one pressure you truly need to understand, and that is "service pressure." Service pressure is the maximum pressure in pounds per square inch (psi) below which you can safely operate your particular shotgun and above which you can anticipate trouble. SAAMI organized U.S. gun manufacturers to adhere to a set of standards for service pressure, and all modern gun models made in the U.S. are proofed or tested to make sure they withstand these pressures.

Service pressure is predicated on the diameter of the bore (the gun's gauge) and the length of the chamber in inches. It is understood that the standard gives you a tiny bit of leeway in making a mistake in handloading. If you exceed the service pressure with a load delivering 10 or even 100 fps greater than your load data suggests is correct, your load will probably perform just fine and your gun will be safe. For your own sake though, you should consider the service pressure to be an inflexible ceiling.

PRESSURE PROBLEMS

Loss of pressure during the burning cycle diminishes any chance of a complete and efficient powder burn. The result will be shot pellets without the energy or velocity you normally expect, as well as a greater build-up of residue in your barrel.

If you start to see inconsistent pressures and velocities with your handloads with any kind of regularity (you can feel this in the recoil, hear it in the report, and often see this in how your targets are struck), you're getting clues that your load components may be breaking down. Irregularities in the pressure curve, perhaps caused by a component shift, reduce the effectiveness of the propellant burn, because compression momentarily slackens. These unpredictable component shifts have several identifiable origins: weak cushion sections in wads, weak seals,

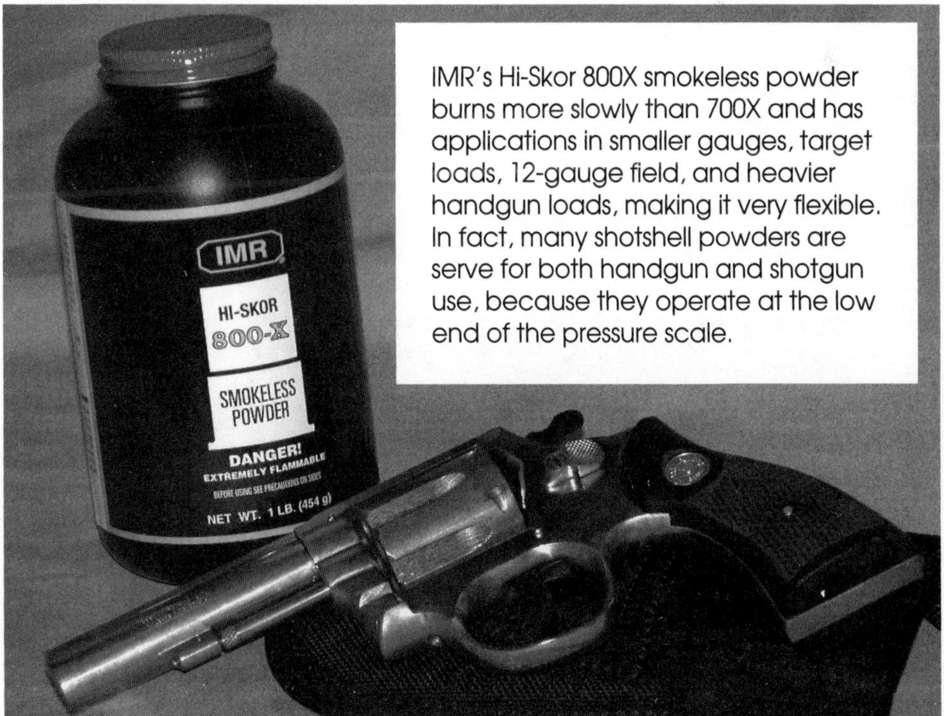

IMR's Hi-Skor 800X smokeless powder burns more slowly than 700X and has applications in smaller gauges, target loads, 12-gauge field, and heavier handgun loads, making it very flexible. In fact, many shotshell powders are serve for both handgun and shotgun use, because they operate at the low end of the pressure scale.

or worn-out hulls with poor crimps. Many substandard loads sound fine when they are fired, but, compared to a more perfect load, produce inferior velocity and energy delivery. If your load experiences a loss of pressure, even a tiny loss, pellets may be getting out there, but not with the speed and energy you expect.

It may surprise you to learn that shotshells are designed to operate at the low end of the pressure spectrum. All powders, particularly those that burn slowly, are sensitive to compression and, with improper or leaky containment, burning will inevitably be less than complete. It is also true that a lighter than average load may reduce your chamber pressure to a point where the proper burning cycle cannot be concluded.

One final note on the subject. When you pull the trigger on a reload and the gun seems to fire, but without its customary, authoritative kick, you may end up with an obstruction in the barrel (likely the spent wad). If this happens, stop shooting immediately and, with the barrel pointed downrange, unload the gun completely and perform a visual inspection of the bore and chamber. If you don't do this and fire another round behind a wad stuck in the barrel, the resulting traffic jam in the bore can cause a permanently ruptured barrel and could hurt you and those around you.

FIRE IN THE HOLE!

When the gun is fired and the powder is ignited, the powder begins combining furiously with its own oxygen to create an ever-expanding column of ultra-hot gas, though this is at a constant pressure (for maximum pellet velocity), and that column of gas takes up an increasing volume of space. This is what pushes the shot out toward the target. The

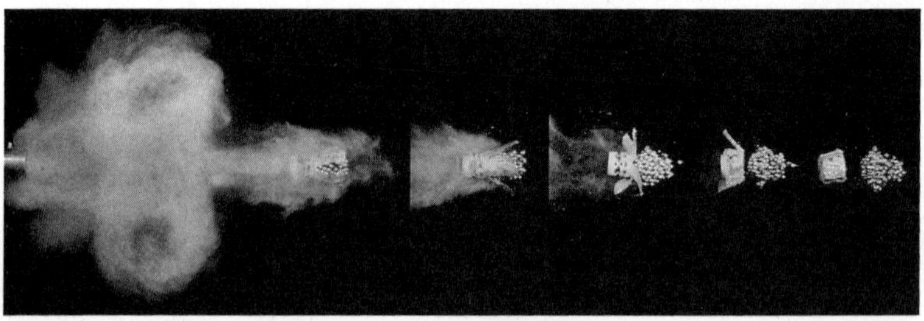

This high-speed photo shows the progression of shot and wad from the muzzle of the barrel to just a few feet in front of it. You can see how quickly the wad falls away from the shot, as we well as how grouped together the shot charge is during its first few feet of travel through the air.

column of hot gas is engineered to balance the burning powder with the expanding volume, hence, powders are grouped by "burn rates." Here are the generally accepted guidelines for burn rates:

- Lightweight loads with moderate velocity, typically target and small-game loads, fall into the faster burning powder category. (Example: Hodgdon's Clays.)

- Heavier hunting loads, which typically have greater velocities, predominantly use slow-burning powders. (Example: Alliant's Herco.)

- You may have to adjust your load for your shooting conditions. Many hunters endure the wet and very cold weather. Cold both retards ignition and causes powder to burn more slowly. To optimize your shotgun's performance in these conditions, you might need specially constructed loads, perhaps some using a medium burn rate powder. You may also want to consider using heavier loads and new hulls that can hold stronger crimps. Also, hot primer brands are more likely to produce good powder ignition in cold weather, because they were developed for just such a situation.

- Pressures for the most effective velocity and kinetic energy must be different for lead than they are for steel or other types of shot. Lead shot absorbs energy more readily than steel shot, which is a third lighter than lead. Therefore, steel normally must be shot at a higher speed in order to achieve the equivalent killing power of lead.

- The more kick to the wad you provide with a hotter and heavier load of propellant, the greater that lead shot pellets will tend to deform and spread out your pattern. Steel and most of the non-toxic shot available for reloading will not deform at any speed. Just remember that, with high speeds, you'll pay the price in increased recoil.

Alliant's Herco is a popular and versatile shotshell reloading powder, along with Green Dot and Red Dot.

POWDER CHARACTERISTICS

In its earliest form, blackpowder was a formulation of about 75 percent saltpeter, 12 percent sulfur, and 13 percent wood charcoal. A proper combination of these ingredients provided a relatively controlled, high-speed burn. One can only imagine that arriving at this "proper combination" took many lives, because improper combinations resulted in dud rounds, uncontrolled, premature explosions, and uncomfortably separated appendages.

If you choose to shoot older firearms, perhaps in events specifically designed for them such as The Vintagers get-togethers, you will want to research powder and shot formulas that will shoot effectively and safely in your older Edwardian or Victorian side-by-side.

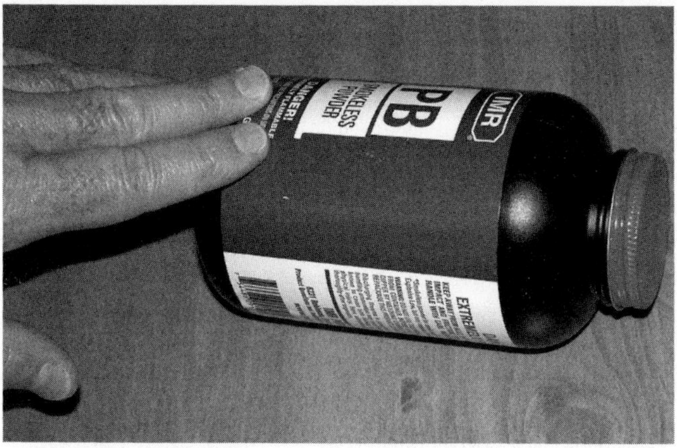

A careful rolling of your bottled powder may help remix it, if it has been handled often.

It can be argued that the smokeless revolution of the nineteenth century turned powder from an explosive into propellant and, thus, fundamentally changed all the known ballistic equations and even the internal structure of firearms and projectiles. Taking advantage of a confined propellant's relative stability and controlled burning properties is the essential characteristic of internal ballistics today.

Smokeless powders are made in three forms. The first is a thin, circular flake or wafer, the second is a cylindrical shape, and the final is a small sphere. Today's smokeless, nitrocellulose-based shotgun powders are also considered to be either single-base or double-base. Nitrocellulose is a pulpy, cotton-like material derived from cellulose (a roughage or fiber such as finely-ground tree trunks), that's treated with specific formulations of sulfuric and nitric acids. Sometimes the powder manufacturer adds nitroglycerin. Single-base powders do not contain nitroglycerin, while double-base powders do.

Nitroglycerin is a heavy, colorless, poisonous, oily, explosive liquid obtained by nitrating—which means to combine with nitrogen—glycerol, a form of alcohol. The well-known nitroglycerin is a fast-decomposing chemical compound with molecules that can be thought of as tightly wound springs. Apparently, it just sits there, waiting for a chance to unwind or "deflagrate." Deflagration is a very slow explosion/very fast burn that generates subsonic pressure waves. Contrast this with "detonation," which produces supersonic pressure waves. Thanks to nitroglycerin's tightly wound chemical spring, double-base smokeless powders such as Alliant's Red Dot generate tremendous energy for their volume compared to that of blackpowder, thus, reloaders can use less powder for specific results.

Smokeless powder granules burn only on their surface, as they have plenty of chemically built-in oxygen. Larger granules burn more slowly than smaller granules.

HANDLING POWDER

Unless a bottle of powder specifically recommends against this practice, it's a good idea to slowly roll the powder on its side for a moment or two before using it. The idea is to remix powder that will already have been handled a dozen times, shipped halfway across the country, and may have been sitting on a retail shelf for months. Obviously, remixing is easier with round bottles than with other containers, but a slow and steady turn or two of a square container will accomplish the same thing.

Do not overdo this remixing. If you are overzealous, you can end up breaking apart the granules and reducing their size through friction. This could alter the burn rate of the powder, even making it unsafe to use in certain recipes. Yes, powder can be vibrated hard enough during shipping to break down the granules into finer units and, so, metering of your powder during the actual reloading process may get a little strange, too. The best assurance against problems is to weigh your powder charges and trust your scale.

BURN RATE

A powder's formulation and granular shape define its "character." Inert retarding chemicals that alter the burn speeds and energy levels of powders are mixed with base elements to stabilize each batch or lot of powder. This brings individual lots into a standard performance range associated with that powder. For reloading, we use standardized performance measurements for dropping a specified amount of a particular powder. This means that, when we run out of our favorite powder and order another can, we can expect similar performance without recalibrating the amount of powder every time.

Smokeless powders may roughly be divided into three burn rate categories: fast, medium, and slow. Most often, we refer to the rate of burn, measured in milliseconds—0.000001-second—to determine performance parameters. This characteristic is a powder's "character" or its application.

The burn rate isn't printed on canisters of powder, because of a concern that someone might wish to calculate a load based upon partial information, something best left to ammunition manufacturers that test under highly controlled conditions. It's also not printed because burn rate is relative, depending upon the condition of its use. However, understanding that there are differences in powder burn rates and how pow-

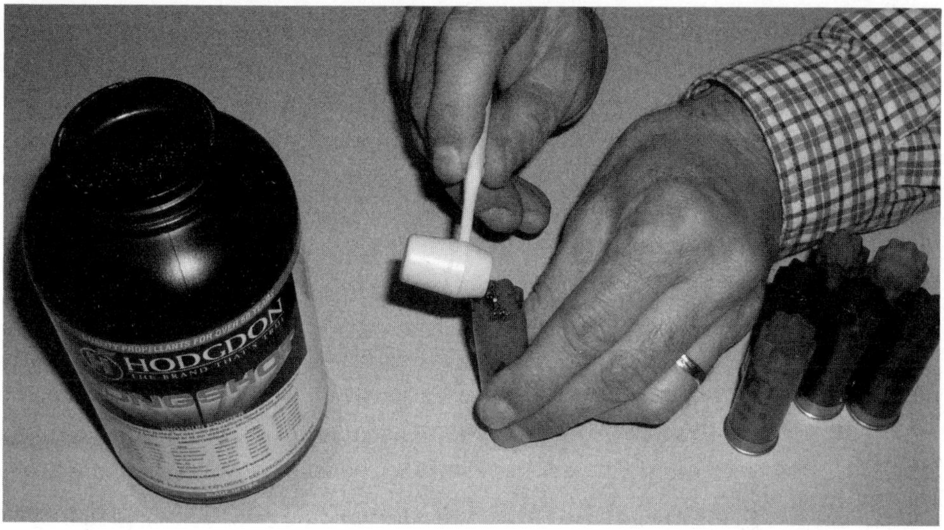

If your reloading press breaks, you can build loads by hand ... with the right measuring devices, that is. A 1-1/8-ounce load of #8s will be about 460 pellets. You can count those. You can also use hand dippers to check the throw of your loads.

ders can be applied is useful information for handloaders. Here are a few basic facts, rules and tips concerning smokeless powder propellants:

- Fast burn rate powders are typically associated with light shot payloads for any gauge, usually in trap, skeet, and sporting clays shells or for small game such as quail and squirrels. Larger shot payload weights used with fast-burning powders would make the pressure skyrocket, with results that could be unfortunate.

- With fast burn powders, it is harder to detect imperfect burns than with slow burn powders. A fast burn covers more defects.

- Everything with powders is relative. "Fast burn" only means faster when compared to other powders that are slower.

- Fast burn powders are less "elastic" in application and forgiveness, compared to slower burn rate powders.

- Fast powders are generally useful in larger gauges. Assuming an equal charge of shot is used, the difference in fast- and slow-burning powders in the sub-gauges is dramatic.

- Smaller gauges put medium and slow burn powders to best use. Smaller chambers generate very high pressures with fast burn powders. Small-chambered guns use relatively slow burn rate powders and usually generate higher chamber pressures than the larger gauges.

- Cold temperatures retard all propellant burn rates to some degree. Slow burn rate powders may not ignite well under severe conditions.

- There is not a parallel relationship between the potential velocity of a load and the relative burn rate of the powder, but the two are related.

- *More powder does not mean additional velocity!* In fact, a drop in velocity, as well as other negative consequences, may occur with too much powder in a shell.

- Light payloads of shot combined with slow burn rates can result in a great amount of unburned powder and fluctuations in pressures.

- Weak crimps, especially in older hulls, or air pockets left in the powder area can result in unburned powder and even "bloopers," where the shot barely dribbles out the end of the barrel.

- Use nominal wad pressure when loading, in other words, just enough to force out all of the air in the hull and securely seat the wad. This is all that is required for proper combustion.

- Store powders only in their original containers. These containers are designed to burst at very low pressures, which only allows the propellant to burn quickly, rather than explode. Solid containers could create, for all intents and purposes, their own bombs. Your propellant does not know the difference. Keep containers tightly sealed and stored in a cool, dry place.

- Never put one type of powder in another type of powder container and never mix powder types.

- Be familiar with the ordinances pertaining to home storage of powder and the amounts allowable in your town.

POWDER ODDS AND ENDS

Most reloaders want a "clean burning" powder, but such a thing depends on how and when you use a specific powder. Some powders, for instance, leave more unburned residue clinging to the barrel than others, but a small change in environmental conditions might reverse those results.

Any head-to-head comparison of powders at the hotter end of the burn rate spectrum (high energy and low volume) against slower burning powders cannot be very scientific, because it ignores application. Faster burning types, for instance do not have as high a percentage of the inert retardants as do slower burning powders. Thus, a slow-

Now part of Hodgdon, IMR is the oldest powder manufacturer in the U.S. IMR's Hi-Skor 700X is a double-base shotgun powder developed for 12-gauge target loads, but it also has applications in handgun rounds.

burning magnum powder leaves substantial residue in your barrel—but you don't want to use a target powder for a high-speed magnum load, especially for something as trivial as having a "clean" barrel. What I'm getting at here is that you need to buy powder to meet your shooting needs and your expected results. It's just a bonus if you powder ends up being "clean."

Shotshell powders have wide variations in ignition *quality*. As I have mentioned, in freezing temperatures, one powder may be harder to set off than others. Some smokeless powders operate better in one gauge than in others, too, and the manufacturers are generally very good about testing and noting this fact. Powder ignition character is altered by other factors, in addition to temperature, humidity, and gauge. Certain types of hulls, for example, are better hosts for specific powders. The sealing quality of a wad, its compatibility with the hull, and a recipe's shot payload weight will also alter the power and timing of ignition, as do the crimp and hull condition. For the new handloader, any ignition problem can be frustrating. Usually, the source is a small problem of misapplied components.

Some powders are bulky and occupy a large amount of hull space. Powder volume has a great influence on the wad column height and which wad fits best. This factor operates both for and against us in handloading, because it may either give or take away component options.

Powder is measured by weight in handloading, not volume. How does that affect your loading? Let's say you want to use compression-formed hulls with a single-base powder and a heavy shot payload. If that's your choice, you will truly need to crush your loads closed. Some would argue that data charts that showed both volume and weight measurements would solve that problem, but it really wouldn't. As a handloader, it is better to become familiar with different powder types and how they occupy space, as well as their energy levels, relative to other powder types.

A lot of that familiarity will come straight from your loading data manual. Pick a page for a particular hull and load for your application, say clay target shooting, and notice the expected velocities for a given set of powder charge weights and shot payloads. It should be easy to see when a powder, primer, and/or wad change is needed to achieve the desired result. In fact, one look at a page of your reloading data manual should make it obvious that there are so many variations, so many factors that affect how your powder works, that experimentation on your part is not the way to go. Always follow tested load recipes and never substitute components.

THE "PERFECT" LOAD

It is often written that new reloaders yearn for a "universal load"—one powder, one weight, one component, and all of it easy to load and a dream to shoot. That scenario is far too idyllic. We all might like the perfect load, but even if loads were so simplified, the shooting conditions would not be. Game bird hunters, for example, have many different ammunition needs. Weather conditions impose additional hurdles and, if we handloaders are to become super-specialized for best possible results, we want to fire refined shells specifically designed for the game, weather, and type of gun used. Powder has a great influence on the overall picture.

Most powders are developed for a specific purpose and then are put to work in other applications. For instance, the old "SR" in IMR's SR-7625 powder stood for "Small Rifle." Still, this propellant worked very well for many shotshell applications. Indeed, shotguns and handguns actually share numerous powder types, but even shotgun-specific powders are stretched in different directions. Alliant's Blue Dot is fine for magnum loads of lead or steel, but it loses its performance edge below 32 degrees. That's just one example, and there are many more. The beauty of shooting a shotgun is the wide world of possibilities in which to use one, and each opportunity has loads that work best for it under specific conditions.

For a powder to have great "elasticity"—to have many uses—it must exhibit good ignition through a wide range of loads with no sharp spikes in pressure. Yet, when stretched over a wide range of loading applications, some powders will reveal qualities that reduce their usefulness or efficiency. It is knowledge of powders and their individual characteristics that allow handloaders to make better load choices.

RECOIL

Recoil is unavoidable, but, as a reloader, you are in a unique position to manage recoil and tailor your loads to your own shooting comfort preferences.

I was reminded of the effect of recoil on a sporting clays course not long ago, when I needed an extra shell at the last station. A buddy fished in his pouch and handed me a 12-gauge load that I automatically popped into the bottom chamber of my Weatherby Orion SSC over/under. I nicely clipped the first of a report pair, a 25-yard left to right crossing bird, with a $1^{1}/_{8}$-ounce Wolf load of No. 8 shot. When I pulled the trigger on the second bird, a longer range target that was zipping quickly away on a flat trajectory, the recoil hit my shoulder and cheek like a hammer. That second shell had been a magnum load of No. 6 (why he had a field load with him on a sporting clays course I'll never know), and it carried more than twice the recoil energy of my first shot. Fortunately for my friend, he could duck and run faster than I could recover and sprint after him.

There are numerous ways to handle recoil. There are soft cheek pads and thick, after-market butt pads for the gun itself, the newer varieties filled with gel-like substances. You can even add weight to your gun with things like mercury recoil suppressors stock inserts (an option more in favor with clay shooters, who either walk very short distances or use carts between shooting stations, than with hunters who may walk miles in a day), because a heavier gun absorbs greater recoil energy than does a lighter gun. Some shooters choose a semi-automatic shotgun, because they'll dissipate a little recoil energy before it reaches your shoulder. Other shooters have their barrel's forcing cone lengthened and the barrel ported; both are pro-active steps that are said to reduce recoil (and improve pattern performance). The barrel porting certainly helps control muzzle rise, and this makes any second shot smoother.

All shooters experience recoil, but all experience it differently. This is because actual recoil is different from felt or perceived recoil. Actual recoil can be determined mathematically. It is a precise, calculated energy transfer. Felt recoil cannot be determined mathematically, because it is completely personal in its effect. Felt recoil is what you experience and tell others about. It depends on adaptations to your shotgun (a thick recoil pad, for instance); the weight of your gun; your body shape and size (large, muscular people will often report less recoil from any specific load than will smaller, leaner people); and possibly even your mood at the time you take the shot.

Whether you are a weight lifter or a professional shooter, you must expect recoil and must have a plan to deal with it. If you do not, it will manage you. Even experienced shooters are subject to flinching, as their bodies learn to anticipate the shot and their muscles tighten prematurely. This can become so severe that it results in full-blown target panic, where the shooter literally cannot pull the trigger.

If anticipating recoil gets to be a problem and you have already worked with after-market recoil-reducing devices, you can look to two options to better deal with it. The first is to find a shooting coach who specializes in gun fitting. Sometimes a poorly fitting gun stock will hammer you severely. One adjusted correctly for personal dimensions of length of pull, cast-off/on, and pitch will help your body deal with recoil much better. The second is to think about reducing the weight of your shot payloads. A good clay shooter, for example, who drops from $1^1/_8$-ounce to 1-ounce loads will probably not notice a difference in scores at the end of the day, but they will notice their discomfort level while shooting dropping. Study the load tables in your loading manuals for loads marked "light recoil," if you find a flinch overtaking your shooting pleasure.

16 Understanding Shot

Lead has been the traditional material for shot virtually since the beginning, with the invention or discovery of saltpeter as the basis of blackpowder. A naturally occurring element, lead is relatively soft, malleable, and abundant. It is also heavy. Its molecular structure is amazingly dense. This means that lead will absorb and carry energy very effectively, so, if you can also make it fly effectively, it works superbly as a projectile. Today's shot options, of course, are much wider than just lead, but, since it is still the most prolific shot type and the type you'll most likely be loading the most often, it's important to understand it.

HOW LEAD SHOT IS MADE

The first step in building lead balls for shotshells is to haul lead ingots to the top of a 100-plus-foot tall "drop tower," where they are then melted and poured through a sieve. The size of the holes in the sieve correspond to the shot size needed. As the molten lead drops, in a mostly round shape, into a vat of water at the base of the tower, it cools and hardens. An alternate manufacturing method called the "Bleimeister process" uses a very short drop of about one meter into hot water for lead or bismuth, thus miniaturizing the shot building operation.

Today's standard options for shot that will successfully tackle any shooting task are wide and not only include size (from tiny birdshot and #9 to hefty buckshot pellets), but composition as well (lead, steel, bismuth and various tungsten mixtures).

After the lead droplets cool and harden sufficiently, they are rolled down a series of carefully spaced plates in a quality control measure that you can think of as a "roundness test." Round pellets are propelled past carefully measured gaps between the plates by the force of gravity. Out-of-round droplets fall through the gaps and are re-melted. Very little goes to waste in this process. The almost perfectly round pellets are then usually coated with graphite. This reduces the surface friction when they are in contact with other pellets or are poured through measuring instruments.

Traditionally, lead pellets have contained both arsenic and antimony. Arsenic is valuable, because a tiny amount, about one-half of a percent of a shot pellet's total volume, increases a lead droplet's surface tension. This makes the resulting pellet harder. Antimony is a non-toxic element that, at a minimum, is used at about the same level as arsenic, and for the same purpose: to increase surface tension. This increases a lead droplet's propensity to cohere as a sphere, rather than some oddly shaped glob. Standard dropped shot has as little as one percent antimony, chilled shot (shot that is cooled as it falls), often contains three to four percent antimony, and hard and magnum shot contains five to six percent.

ROUNDNESS

Fact: round does not fly well.

Aerodynamics is a human endeavor based on nature's principles. These principles suggest there are several shapes (oblong and spherical, for instance) more efficient as projectiles than spheres. However, to produce other shapes and sizes in the enormous quantities and at the low cost needed to supply the world of shotgunning, and then orient such shapes correctly in the shotshell for best flight characteristics, would require a whole new (read "expensive") approach to shotshell design and materials processing. Yes, there are companies experimenting with shot design that's other than round, including Winchester with its Blind Side shot, Federal Ammunition and its Black Cloud shot, and a private indi-

All things considered, round is the perfect shape for reloading projectiles. Round pellets are cheap and relatively easy to produce, and they are easier for the reloader to load than, say, oblong-shaped pellets. Additionally, if they are not deformed at setback or in their travel through your barrel, they fly true.

vidual that created a shot he called Squound, among others. Winchester and Federal have found outstanding success with these loads and their intended market of waterfowl hunters, but, since non-round shot types are not readily available to the reloader and can reqire special processes for reloading correctly, they must be considred an advanced reloading topic. For the purposes of this book, we'll stick to round shot, just know that there are other types out there.

Although it is easier to push a needle through almost any solid matter than it is a marble or a small BB, spherical pellets have proven for years that they can get the job done. If pellets remain perfectly spherical and pressure on the pellet remains proportional around it, the pellet flies true and straight and with a load of energy for penetration. If a pellet becomes deformed during the firing process, perhaps because of the crushing setback when the powder ignites, because the wad is defective, or because the shot scrapes against the barrel, air pressure will not be proportionally distributed around that pellet. Scrapes, dents, and flat spots on pellets cause them to make continual mid-course corrections, and this is not usually in the direction of your intended target.

Deformed pellets waste energy and become fliers, ending up behind and on the outside fringe of a pattern. Because softer pellets deform more than harder, high-antimony magnum pellets, their pattern is usually wider. While this is fine for close targets and skeet shooting (indeed, it might be excellent), an extra-wide pattern often causes you to miss a duck or a sporting clays target crossing at 45 to 50 yards. It would not be an appropriate trap load either, because even a Full choke cannot fully compensate for the spreading ability of soft, cheap shot.

Nickel-plated lead shot is available in sizes from No. 2 to No. 9. The non-porous nickel plating process produces hard round pellets that hold their energy longer and penetrate better because they have less deformation than standard lead shot. Nickel-plated shot may have a slightly larger size variation due to the heavy plating process and reloaders should expect that the number of shot in a load may vary if this shot is used.

PLATING

Lead shot is available with a thin coating of either copper or nickel. While copper is not much harder than lead, it was the first metal used extensively to "hard-coat" lead, perhaps because it was easy to work with.

Winchester developed Lubaloy copper-coated shot, in 1929, and marketed it aggressively to hunters on the theory that harder pellets held a much better pattern, because they retained their spherical shape better. The company claimed that copper prevented lead from fusing and was also more "slippery" than lead, thereby promoting improved pellet flow through the barrel. Still, even though many thousands of hunters swear by it, there is only anecdotal evidence that copper improves lead shot performance. In fact, copper may be nothing more than a marketing ploy, a thin coating that makes shot "prettier," rather than actually increasing its functionality.

Nickel plating, on the other hand, *does* change the hardness of shot, even though, like copper, it glazes pellets with only a very thin coating. Some writers have referred to nickel-plated lead shot as the "ultimate game pellet," because of its increased energy transfer and, hence, penetrating ability.

Steel shot is often coated with copper or a zinc alloy. The purpose of coating the steel is a little different from that for coating lead. Exposed to the slightest moisture content, steel shot quickly rusts when it is contained inside a paper or plastic shell, so coatings, in this case, act as rust inhibitors.

Many reloaders believe that spraying their pellets with a dry lubricant helps them crowd through a choke and exit the barrel a bit easier. This practice actually cannot hurt. Although the original manufacturers coat most lead shot with a graphite powder, most pellets, regardless their material makeup, may actually benefit from such an application.

BUFFERING

Adding buffer to a load fills the spaces between the shot pellets with a lightweight material that has a long shelf life. This buffering material is supposed to soften the effects of setback force, preventing it from exerting a crushing impact on and changing the shape of soft pellets. The perfect buffer will also carry its cushioning action through the forcing cone and up into the choke.

The open area between the pellets defines the amount of buffer that can be placed in a load. Since the amount of space varies when different pellet sizes are used, it is impossible to state, in general terms, the exact amount that will be needed. Other factors, such as the amount of shot in the load, the density of the buffer material, and the method used by the handloader to insert the buffering material, also come into play.

Because so many elements are involved—air temperature, primer choice, etc.—the pressure differences generated in certain loads by adding a buffer do not follow a general rule. Some loads reflect only very small changes when buffered, while others run high. To that end, you should buffer only those loads that have been tested with buffer per the printed reloading data you're using; always assume a rise in chamber pressure whenever adding mass to any load.

A note of caution: Some reloaders have used finely grained substances like flour, salt, and cereal products (dry, uncooked Cream of Wheat is a perennial favorite). Actually, flour can be entirely suitable if it is shot immediately on a dry day, but flour tends to absorb moisture, pack down hard, and increase the pressure curve. Singed flour inside your barrel requires extra brushing before the solvent patch can be run through. Because its use is so narrowly defined, I cannot ever recommend its use, nor any of the other substances that aren't purpose-specific, load data-recommended buffers, because it can ruin your gun.

What about buffering steel shot? The resiliency of steel causes its pellets to bounce, when the powder ignites. Energy is exchanged between pellets, rather than absorbed through deformation as happens with lead shot. Steel shot buffers contain ingredients that deaden these interactions, cushion the shot, and reduce the wave of energy passing through the pellets during the initial thrust of a burn cycle. Most steel shot loads are buffered, since the advantages of doing so are essential for high performance.

There's another reason to buffer steel loads, though not with a granulated buffering media. The larger pellet sizes found in many steel loads, such as T or F shot, sometimes do not disperse evenly in the base of a shot cup, due to the wad's interior space constriction. Thus, one pellet may occasionally support the strain of the entire load. When this happens, it places extreme pressure on the base of the wad. When using large pellets (T, F, BBB, and even BB), consider placing a thin wafer of cardboard or wool felt in the base of the wad to act as a reinforcing cushion, if there is room. This will allow the pressure of setback to spread over a wider area and ease the strain on the base of the wad. Additionally, some reloaders find that placing an overshot card wad on top of a steel shot load will better hold pellets within the shotcup for the journey up the barrel. This reduces the possibility of a pellet getting out ahead of the wad and scouring the barrel. Any card wad of thickness .030-inch and less will serve the purpose, and it will get out of the load's way upon exiting the muzzle.

OTHER NON-TOXIC SHOT

It's been known for many years that lead is a poisonous. Historians speculate that the fall of Rome was due, in part, to the prevalence of lead pipes used to carry drinking water to the cities of that empire. (Incredibly, some of these lead pipes are still in use, bearing the imperial seal of a Caesar!) In humans or animals, lead poisoning isn't pretty. It induces muscular weakness, cold sweats, vertigo, and vomiting of blood. But unlike a vial of arsenic, lead's debilitating effects occur slowly, though they will inexorably result in a miserable death.

When lead was banned for the hunting of migratory waterfowl, steel was the only commercially available alternative to lead, but it almost immediately proved problematic. A treated alloy of iron, steel is harder, but one-third lighter than lead and only 65 percent as dense. The flight characteristics and energy transfer properties of steel shot (i.e., penetration), were inferior to that of conventional lead shot. Steel accelerated quicker to muzzle velocity, but it also slowed down significantly faster. Hunters hated it. Several writers credit steel shot for the return of the 10-gauge. Their argument was that large steel pellets, even polished steel pellets, simply don't flow very smoothly through smaller bores, not even the 12-gauge.

Today, steel shot not only rivals the lead loads of yesterday, it often surpasses its performance, thanks to innovative wads and improved, purpose-specific powders (thanks in large part, to the success of popular, highly engineered factory shells like Remington's HyperSonic steel, Kent's Fasteel, and Federal's Blind Side waterfowl). In

Not only has an intense study of steel as an effective waterfowl pellet been completed during the past 20 years, but powder companies have responded with safe and effective propellants for the new steel. Today, many students of shotgun ballistics believe that with careful selection, you can build steel loads that are just as effective as lead loads.

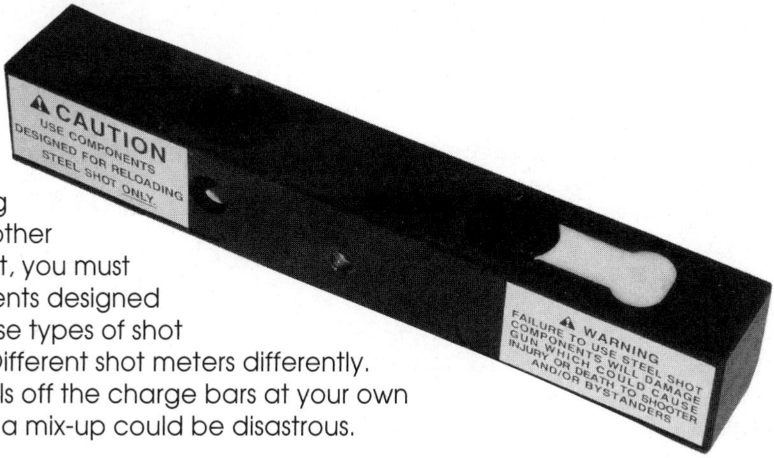

When loading with steel or other non-toxic shot, you must use components designed to reload those types of shot specifically. Different shot meters differently. Peel the labels off the charge bars at your own risk, because a mix-up could be disastrous.

addition to components and data that allow the reproduction of these super-steel loads, consumers also have a widening variety of other non-toxic shot offerings, many much heavier than steel, and some including shot that is other than round in shape. Though it's lost much of its original favor among hunters and recreational shooters, bismuth, the original alternative to steel, is still a reloader's non-toxic alternative. Thankfully, it's now joined by a number of other "designer" non-toxic shots, such as Nice Shot produced by the company Ecotungsten, Hevi-Shot from EnvironMetal, a development called American E-Shot (available at www.reloadingspecialtiesinc.com), and TomBob Outdoor's ITX Shot (available at www.ballisticproductsinc), to name a few.

17 Shotshell Reloading Presses

For a few hundred dollars, sometimes less, you can be set up with excellent new equipment and buy all the components required to begin building your own shotshells. You will need a place to set up your press, preferably a bench setup that allows you enough room to set out your components, has a place for your finished cartridges to reside, and permits you enough room to comfortably move the ram arm of the press through it's full range of motion. It is also helpful to have your setup in a place where you will not be continually distracted and where you can be reasonably certain that other hands, especially those of children, will not meddle in the components.

Shotshell presses come in two forms, single-stage and progressive. I strongly recommend that, if you have never before loaded a shotshell round, and especially if you don't have an experienced reloader friend to help you learn, that you begin with a single-stage press. Such a press permits the loading of exactly one shell at a time, one step of the reloading process at a time, and its very simplicity will help you learn the feel of primer and wad seating and hull crimping. It will be most beneficial to have a press that sizes shells automatically, as well as one that will automatically feed the primers, which can help eliminate how much you have to handle them. One of the best features of a single-stage press, other than their lower cost, is that they will last darn near forever.

At some time in your shotgunning life, you may become deeply involved with one of the clay sports. When that happens, you're going to want to speed up the reloading process. That means a progressive press. More complex than a single-stage, the progressive performs all the essentials of primer removal, primer seating, wad seating, powder and shot charging, and crimping with one pull of the press handle via a rotating turret that aligns a hull in turn in each of the several stations that perform these jobs. All you usually have to do is put in a fresh hull at the first station each time and start a wad (by hand) in a re-primed and powder-charged hull at the appropriate station. Not hard to see where this is a much faster way to load than with a single-stage press. Let's take a look at the primary brands and types on the market today.

DILLON PRECISION

This company is primarily dedicated to metallic cartridge reloaders, but it took many of the innovative features it introduced in those presses and incorporated them into the SL900 for shotgunners. This one is automated right down to an optional automated hull-feed system. Features like a tip-out wad guide and an extremely accurate powder drop design (there's also an optional low-powder alarm), eliminate much of the hassle users of other progressive presses often encounter. With its oversized top hoppers, you'll need some room for this big press, but, if you have a need for speed that doesn't compromise individual shell quality, then this press—available in 12-, 20-, and 28-gauge—might be just what you're looking for.

Dillon's SL 900 features hull-activated powder and shot systems to help eliminate troublesome bushing changes, as well as spilled powder and shot. According to Dillon, the built-in adjustable powder measure is good to within one-tenth of a grain, and the shot hopper holds 25 pounds!

HORNADY

Hornady makes a wide variety of presses and ammunition, and that's included several excellent progressive shotshell presses. However, as this book was being assembled, the company suspended production of its lineup of Model 336 progressive shotshell presses. Why? As Hornady explains it, in 2013, and like so many other companies in the reloading and ammunition business, it experienced sky-high demand for its products, thanks to the increasing political tensions surrounding gun control. Rather than place items on indefinite backorder, Hornady chose to consolidate some of its manufacturing processes, thus temporarily suspending production of certain products that didn't have the highest demand. Unfortunately, included in that list was the lineup of Loader 366 shotshell presses. Hornady's website statements assure consumers that these production halts are not permanent, so, if this is one of your preferred brands, we strongly recommend you check back with Hornady periodically until production resumes.

LEE PRECISION

Lee, like Dillon, primarily focuses its business on metallic cartridge reloaders, but its Load-All is a uniquely designed single-stage press that is perfect for the new shotshell reloader. Not only is it compact in size, which means you don't need to clear a six-foot-square space in the garage to get going, its linear arrangement makes this one of the fastest single-stage presses you'll ever likely encounter. Available in 12-, 16-, and 20-gauge setups, the Load-All comes with two-dozen shot and powder bushings, providing lots of loading options for the novice reloader without having to purchase numerous additions. Here's the real kicker—any Load-All in any of its three gauges sells for under $75 dollars, so, if you're fascinated by reloading but not sure how deep you want to make your investment, this is a hard price point to resist.

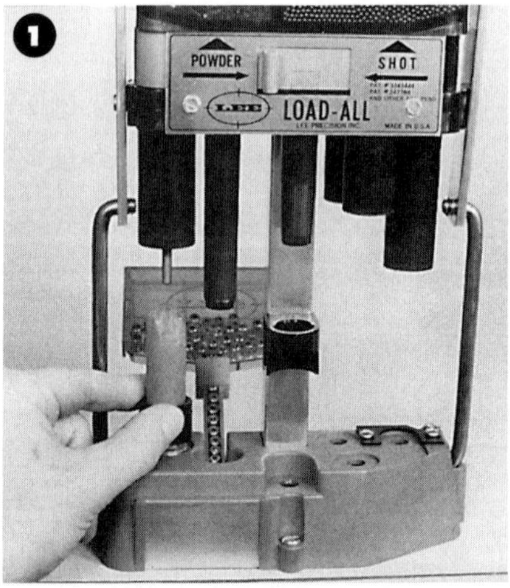

Insert a hulll for de-priming.

The popular Lee Load All II (left) is easy to use and has a legion of proponents. The numbered illustrations demonstrate its simplicity.

2

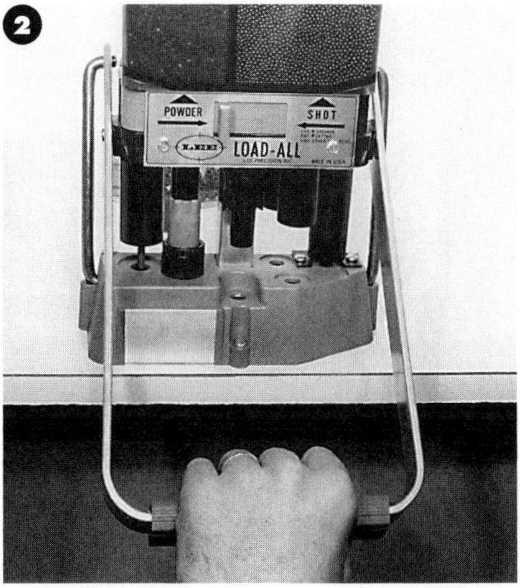

Lower the handle/ram and re-prime the shell.

3

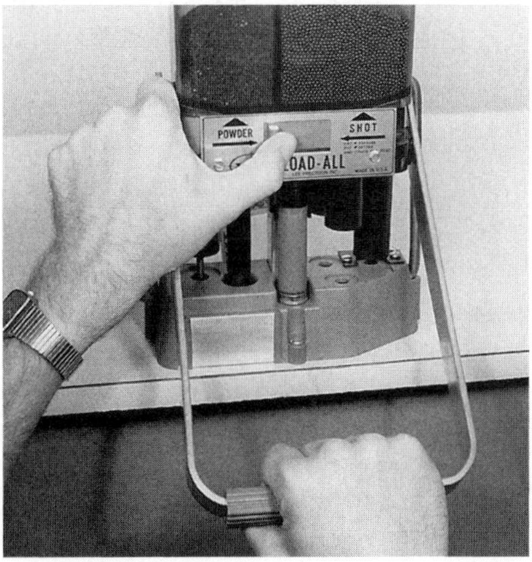

Move the freshly primed hull to the next station and drop the powder.

4

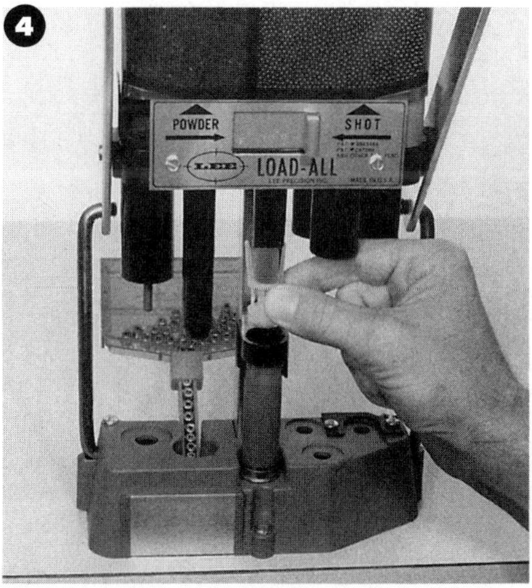

After the powder drop, insert new wad/shotcup.

5

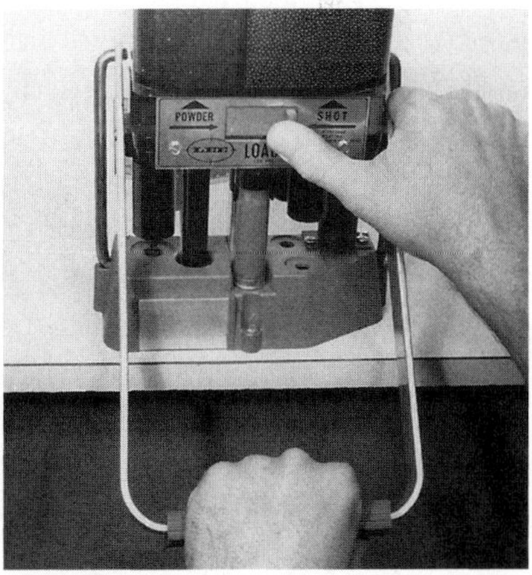

Drop the shot charge.

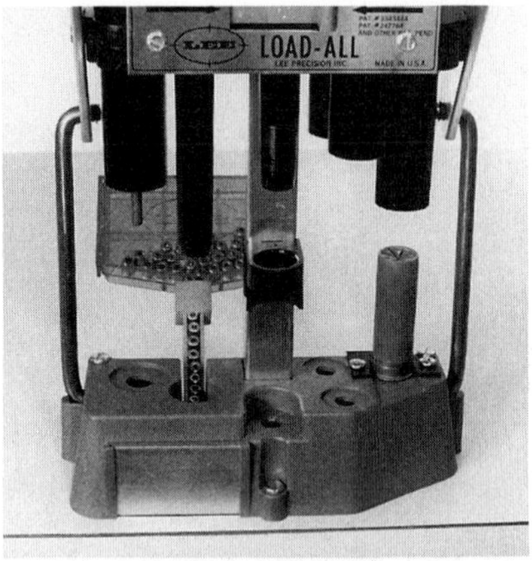

Move the shell to the next station and crimp the shell.

And here's your finished shell!

MAYVILLE ENGINEERING COMPANY (MEC)

Mayville Engineering has manufactured shotshell reloaders under the MEC name since 1956. Everything from the least expensive single-stage machine to a lineup of automated progressive loaders is available, including a press dedicated to the special concerns of reloading with steel shot. There are also machines designed for the challenges of loading slugs, and there are both hydraulic and electric add-ons that really speed up the operation of the progressive presses.

One of the very first machines many reloaders use, the 600 Jr. Mark V, was introduced in 1985. Today, it costs less than $200 for any press under 10-gauge; the 10-gauge press is running just a smidgen under $210 at this writing. MEC says that once the operator gains a little experience, this single-stage reloader can fill up to 10 boxes of 25 shells an hour. MEC believes that its Sizemaster reloader, the next single-stage press in the lineup, is an excellent choice for hunters. The built-in Power Ring Collet Resizer returns all types of shells—brass or steel bases, high base or low—to factory specifications. The Sizemaster is adjustable for 3-inch shells and fills all gauges and the .410-bore. Finally, there are two specialty presses in single-stage design. The 600 Slugger is a single-stage designed just for rifled and saboted 12- and 20-gauge slugs, with the press designed to fold the hull top to look like a roll crimp, just as you'd find in factory loaded rounds. The Steelmaster, which will also work with lead but was designed primarily for steel loads, comes in three iterations: 12-gauge 2¾/3-inch, 12-gauge 3½-inch, and 10-gauge.

The MEC 650N is advertised as "maximum effect for minimum effort." Though it works on six shells at once and finishes a shell with each pull of the handle, the 650N does not resize hulls; it's kind of a hybrid between a progressive and a single-stage. With this press, resizing becomes a separate operation. MEC says that the 650N is "the ideal press for the person who likes to resize and inspect their shells as a separate operation."

Moving on to progressives, MEC's progressive 8567 Grabber mechanically programs 10 operations at six stations. It has a fully automatic primer feed, auto-cycle charging, and the three-stage crimp. The built-in Power Ring resizer operates without interrupting the reload-

The introductory single-stage press from MEC is the 600 Jr. Mark V (above). It can load up to 10 boxes of 25 shells per hour and can be upgraded with an automatic primer feed that eliminates the need to handle each primer individually.

Above, a close-up of a MEC rig filling up a hull with shot.

A good quality progressive press that is operated by a hydraulic unit, such as with the MEC 9000HN, can produce finished shells with every tap of the toe. This set-up costs between $1,280 and $1,296, depending on gauge.

ing sequence. Stepping up, the 9000 Series is MEC's top-of-the-line progressive, offering the best in speed and efficiency with each stroke of the handle, including auto-ejection of the finished round, plus the option to add either a foot-operated hydraulic or a finger push-button operation automator.

The 10 different operations at six stations produce a finished shell with every pull of the handle. This MEC 8567N Grabber has a fully automatic primer feed, as well as auto-cycle charging and three-stage crimping.

PONSNESS/WARREN

Ponsness/Warren (www.reloaders.com) has developed reloading gear for almost 40 years and its Patriot press is state of the art for today's competitive target shooter. Available in 12-, 20-, and 28-gauge renditions, these extremely high-quality progressive presses house features such as hull-activated shot and powder drops, an adjustable shell seating post, and the capability of loading both lead and steel shot. If this is a little much for you to start out with (with a price tag of more than $1,200 dollars, this is considered a serious press for serious reloaders), then the company's New 800 Plus provides many of the same features, can be operated both manually and automatically, but is several hundred dollars less than the premium Patriot model. This is also a good

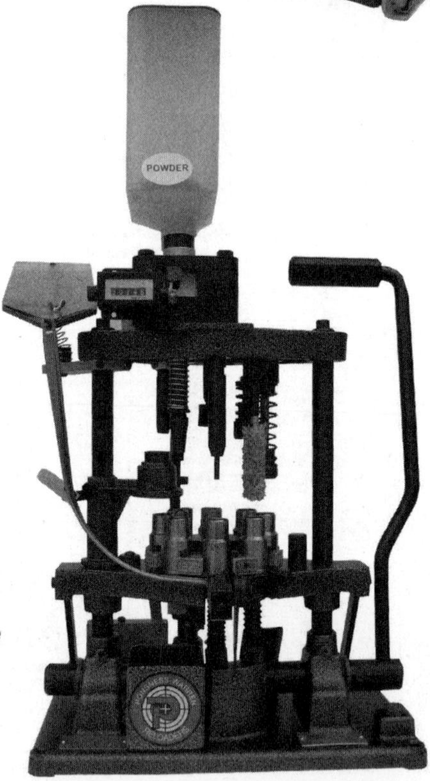

Ponsness/Warren progressive presses don't come cheap. They are one of the primary choices for serious competitive shooters who shoot large volumes in both practice and tournament and need a press to keep up with them.

choice for the avid, all-around shooter, with additional setups available in 16-gauge and .410-bore.

Ponsness/Warren makes a third progressive press, one designed to address the increasing demand for non-toxic shot reloading options. The L/S-1000 requires no conversion kit to convert itself from lead to steel or bismuth and back again. Ideal for the ward-warrior hunter and competitor who encounters public hunting lands and clays ranges that mandate non-toxic shot, this sleekly operating press has P/W's Uni-Drop system, which easily drops shot sizes up to and including BB and works for 12-gauge 2¾-inch, 12-gauge 3-inch, and 10-gauge 3½-inch shells—now *that's* versatility!

RCBS

Focusing only on progressive loaders, RCBS produces The Grand in both 12-gauge and 20-gauge, with conversion kits to the other gauge for both. The eight-stage press is geared for high-volume shooters of all type, with oversized hoppers for both powder and shot. It's also one of the sleekest progressives on the market, with case-activated shot and powder stations that won't allow for so much as a single pellet or powder grain coming down if there's no shell in the station. The Grand also provides unique draining systems for both shot and powder hoppers, making cleanup a snap when you're done reloading for the day.

The Grand is RCBS' top-of-the-line progressive shotshell reloader and a very fast press. Notice this reloader has setup to maximize the efficiency of his reloading process, with a bin of empty hulls to the left of the press and a bin of fresh wads hanging from the bench on the right. The plastic cup with the white lid at the bottom of the press captures spent primers expelled during depriming in the first stage.

This is a big, tall press, one for which you'll want ample bench space (especially if you opt for the additional hydraulic unit, which can push production speed to more than 600 rounds an hour for the practiced reloader). If you're loading for quantity, that kind of space is necessary anyway, for storage of both finished product and components.

SPOLAR

Talk about a gold standard! The Spolar Gold Premier is, literally, gold-colored, but its fancy looks are really just icing on the cake for what many consider to be the shotgunner's premier reloading press. Precisely accurate loads are completely attainable, thanks to vibrating electronic shot, powder, and primer settling system, and the Gold Premier has one of the most reliable primer feeds known. There's also a low-primer warning system, crucial when you're cranking through rounds at a fast pace. All four of the competition gauges—12-, 20-, 28-gauge, and .410-bore—are available, and dies for each can be changed in less than five minutes, a feature of premium importance to a tournament shooter covering all four bores and shooting for score every weekend. Oh, and the 25-pound shot hopper helps with that kind of volume reloading, too, as does an optional hydraulic unit.

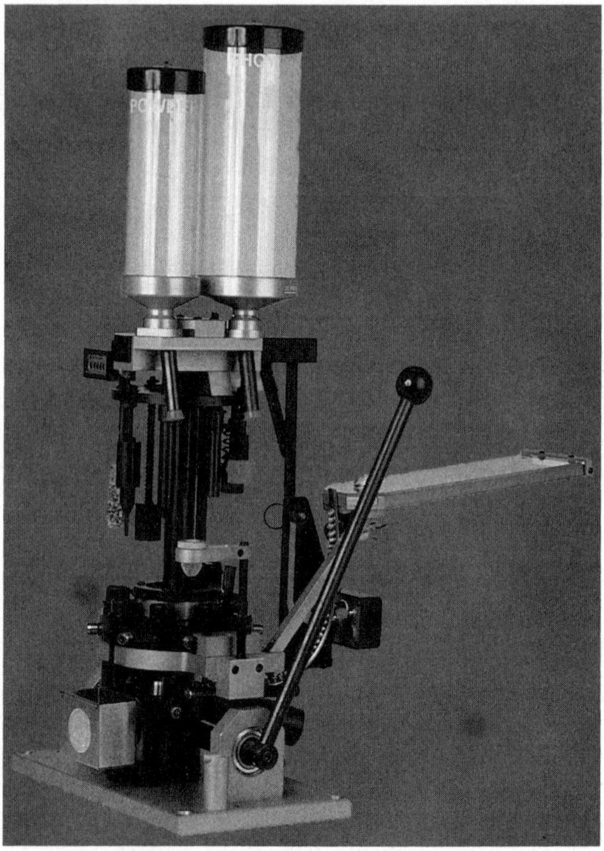

The gold-anodized finish of the Spolar Gold Premier allows this superbly well-thought-of loader to look as good as its reputation. Its vibrating electronic shot, powder and primer settling system is just one of the Gold Premier's features.

SECTION E: Reloading the Shotshell

18 Component Setup and the Reloading Process

Regardless which type press you end up using for the majority of your loading, the steps in the reloading process are the same. Just remember that most of these steps will happen automatically with a progressive press. Also, since the progressive press is the predominant choice among shotgun reloaders, even novices, the individual steps explained in this chapter are mostly geared toward the mechanics of those machines.

COMPONENT SETUP

Prepare your components. Fill your primer dispenser and powder dispenser with the respective brands of each listed in the recipe. Have a basket available to fill with the wad brand and size listed in the recipe. Fill your shot hopper with the size shot you have selected (slugs and buffered loads will be covered in a different section).

Next, you should perform a hull inspection. Inspect and sort your hulls (gauge, then brand, then size if necessary), according to the recipe you've selected to load for. Discard any hulls that are split at the mouth crimps or cracked in the body, that have been squashed and are no longer round, that have damaged brass heads, and those hulls that are excessively dirty either with residue from previous shootings or from range dirt, sand, and mud. Put the hulls that are ready for loading in a basket or clean box that you can grab from easily and set to the left side of your press. Do the same for the wads you'll be using, but set them on the opposite side.

The final steps before you begin loading include preparing the press' shot/powder charge bar (almost all presses have this arrangement; Dillon's SL900 progressive has its powder adjustment at the hopper and is the exception to the charge bar arrangement). The charge bar of most presses has two holes in it. Some are specifically calibrated and labeled for the shot load it will deliver to your shell through a straight, un-beveled hole, while others require a shot bushing to deliver the correct shot charge. The second hole will hold a bushing that meters out the required powder charge. Note: You *must* refer to a bushing chart for your particular press to determine what bushing is needed for the powder you're using and as dictated for the weight of the charge required in the reloading recipe you're following (see the bushing charts in Appendix II). Any one particular bushing will meter out different weights of different powders (though their volume may be the same). Have bushings the next size up or down available to you, in order to adjust the powder charge thrown as necessary during the initial stages of loading. The rule in your head should be measure twice (or more), charge your shotshell once.

Before you begin placing hulls in your supply bin, you must examine each one for significant wear: permanent creases, split crimp edges, cracked bases, or separation between the brass base and plastic hull. Resist the temptation to over-use worn hulls. This can cause loss of pressure in the chamber and barrel and result in dud loads.

THE RELOADING PROCESS

Now that you have your press set up and your components organized and set about your press so that they're easily accessible, you can begin the actual reloading process. Here are the steps in the order they must be performed:

1. As a first step, your reloading press will generally perform two functions at the same time, depriming and resizing. Resizing is necessary because, when you fire a shotshell in your gun, the heat and pressure generated can enlarge the hull slightly, not enough usually to be visible with the naked eye, but enough for a shotgun chamber built to thousandth-inch specifications to resist.

A press resizes a hull in two ways. It can use a steel-fingered sleeve to slide up on the brass and squeeze it or neck it down to the proper size as a single step in the reloading process. A second method is to force the entire hull into a full-length resizing die at station No. 1 of a progressive press. This die can then hold the shell tightly through the entire process of reloading and crimping.

Kurt Fackler, of Ballistic Products, who wrote and edited a former edition of this book, says you may occasionally run into a hull that is so oversized that resizing is impractical. "Rather than possibly damaging your equipment," he advised, "throw the hull into the garbage. If your shotgun is producing these hulls, your chamber may be oversized. It is a gun problem, and you may need a new barrel." This is good advice for any function of reloading. If you find yourself having to force something, remove that hull and discard it and any of its contents properly.

Let's take a look at depriming, better known as removing the spent primer, from your first hull to be reloaded. Again, this is sometimes performed simultaneously with resizing. Unless you are using a hydraulically-operated progressive press, which accomplishes these steps automatically with the tap of a toe, you

Resizing your shotgun hull is a necessary first step in reloading. Here you can see the steel "fingers" at the base of the hull in this press, doing that job. If, for any reason, the hull does not look right or you find the press forcing an operation such as this on your hull, discard that hull.

Step one after resizing the hull. Before you can put a new primer in, you must take the old primer out.

must physically, although indirectly, push the old primer out the bottom of the hull via a pull of the press' lever. This step is gauge-specific, so be sure you have the correct de-priming pin in place.

As it is with resizing, if your de-priming operation requires too much force, check for an obstruction in the hull. Also, the base wad may have shifted. If, on the other hand, the primer seems to pop out too easily, check the hull outside of the base for black deposits, which indicate that gas is escaping past the primer. If that's the case, the hull should be discarded.

De-priming must *not* crush the hull's base wad, as this can cause a drop in pressure for your next load. Also, you should never de-prime live primers from hulls! Pressing a live primer out could cause it to detonate!

2. Re-priming your hull is the second step in the process. Modern shotshell primers have been standardized to the No. 209 size, but, and as has been noted before, just because they're the same size, it doesn't mean they're identical in other capacities. Primers are built with different levels of energy, and each load calls for a specific primer brand in order for that load to function properly. Use *only* the primer brand recommended in your loading data recipe. If you do not have that particular primer, get some or choose a different load.

The fit of the new primer inside the base of a modern hull should not be a problem. If the primer does not fit easily into the empty pocket, do not force it, as, again, a detonating primer will hurt you. Stop and look for the problem and discard a hull that cannot be remedied.

Place a new primer in the priming station, preferably dropped from a primer-dispensing press accessory, but, if you have to, with non-metal

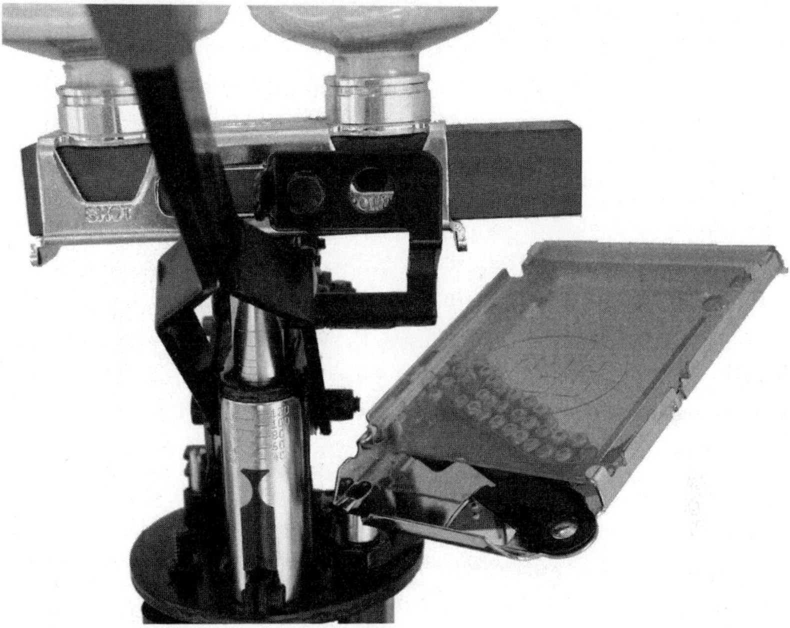

Upgrading your reloader from the most basic functions to include such accessory items (some would say necessities) as an automatic feed tray for tiny shotshell primers usually costs a hundred dollars or so, but over the life of a press it is well worth the expense.

tweezers, so as not to impart the oils of your hand on the priming material. Next, set the hull you've just resized and de-primed back in the die slot and pull the handle on the press down. This will seat the new primer. You should be able to feel the primer seat correctly, but until you're more experienced with the entire reloading process, remove the hull and see that the primer is flush with the brass head.

When the hull has been resized correctly and everything is working correctly on your press, it is easy to seat a primer correctly—and the only correctly seated primer is one that is flush with the bottom of the base. Seat the primer too deep and the hull base can actually be pushed in so that it will not fire or not fire properly. If you experience this with your press, you will have to adjust the primer seating post and begin again with a fresh hull.

If the primer is not fully seated and sticks out, attempt to reprime again. If you are unsuccessful in this second attempt, again, you may need to adjust your primer seating post. Do not be tempted to keep moving forward in the loading process with a primer that is not fully seated. Even if you successfully finish loading the round (possible with a single stage, though more unlikely with a progressive, as the protruding primer will almost likely catch as the turret attempts to rotate and prevent the shell from accessing the next station), you have to know that the shell may not feed through a semi-automatic or pump-action gun. Also, primers extending out beyond the level of the base have, in rare cases, been known to explode prematurely during action cycling before the gun is properly locked. This will cause your shell to, literally, "go ballistic."

Progressive presses will automatically feed the primer into a waiting slot as the turret on the press turns with the upward movement of the press handle after the hull in the de-priming station has had its primer punched out and its brass resized. The primer should drop into the appropriate slot before the de-primed hull, moving in the turret, arrives at the station. You will have to watch to make sure this has happened, and you will also have to watch to make sure the primer has dropped properly—explosive side up and sitting straight in the primer slot. If you see the primer has not dropped explosive side up or is not straight in its slot, take a toothpick and right it before you complete the upward stroke of the press handle. If you see that the primer missed the hole altogether, remove the hull now in the priming station, put a primer in by hand (with non-metal tweezers), and reinsert the waiting hull. Follow the procedures of a single press to seat and check for proper seating of the primer.

Automatic primer feeds can both jam in and of themselves and cause jams in the press. When a primer gets jammed in the primer feeding system, you need to be patient and gentle. Never pry at a stuck primer with a sharp object and don't force the operating handle, because you could cause the stuck primer to explode, and this could cause a chain reaction with the remaining primers or, depending on your setup, with the powder. You will have to disassemble the mechanism carefully to remove the stuck primer. Same goes for a primer that has jammed under a turret. You may have to disassemble some of your press to remove the offender.

A properly seated primer should be flush with the metal head.

Placing a bushing in a charge bar. Because you are reloading your own ammo, it is your responsibility to double-check each bushing, each powder type and your load data. After all, you will probably be the closest person to the chamber if something goes wrong.

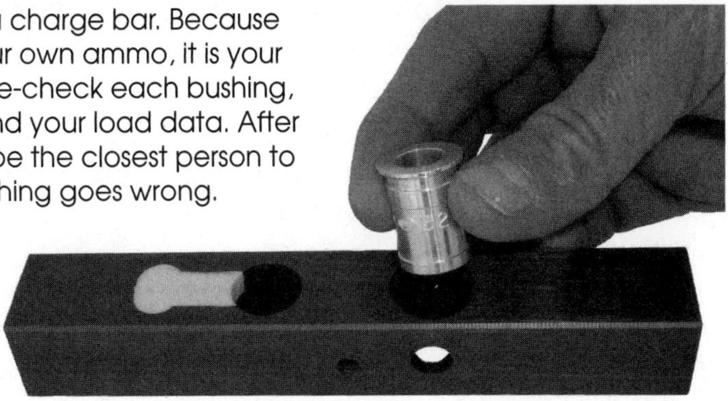

3. Drop your powder charge via the charge bar and powder hopper per the directions for your particular press (charge bars and bushings operate differently on different press brands, and at least one press, Dillon's SL 900, doesn't use a traditional charge bar at all).

Powder is measured in grains. One ounce equals 437.5 grains; 16 ounces, or one pound, equals 7,000 grains. Remember this measure.

Remove the hull from the powder station and carefully pour all the powder into the pan of a properly calibrated and zeroed beam or digital scale (the chapters on equipment for metallic cartridges have excellent advice on scales and how to use them). Does it match the charge required in your loading recipe? If yes, you can move on to the next station. If not, you'll need to pour your powder back into the hopper, reinsert your hull into the charging station, drop the powder charge again, and weigh the charge again. Still doesn't match the recipe? How far is it off? A tenth of a grain, a half a grain, a full grain? Any more than a few tenths of a grain on either side of the recipe load and you should probably explore changing bushings, repeating the process until you are as close as possible to the listed powder charge in the loading data. Alternately, the single-stage shooter can opt to use a powder trickler and weigh each hull's charge by hand, a slow process, but one that gives exact results.

With progressive presses and their full (and heavy) shot and powder hoppers on top, you should weigh the powder charge in each shell moving through the first full rotation of shells. Some progressive presses seem to take a couple cycles to "settle in" (making sure your press is securely attached to your reloading bench and has minimal movement/shake as you operate the ram arm can reduce this potential), and there's less harm to be done in restarting a few shells than there is in severely over- or under-charging a shell. You should also weigh half-way through a batch of, say, a hundred rounds, and again towards the end of your run, as hoppers near the empty state and the overall weight of your press changes. Even with a heavily bolted-down press, there's always *some* movement in a press that affects how its powder meters out (and, to some extent, the shot, too). Too, some presses are just more finicky than others, and it can be good for consistency to top off your shot and powder hoppers when they reach the half-empty mark, while others will deliver rock-solid, on-the-money charges from shell one to shell 300. Powder charge consistency can also vary with the bushing size, the powder itself, and humidity and temperature in your loading area. These are things that only time with your press will let you discover, and until you have several hundred rounds under your belt and have changed loads and bushings a few dozen times, it pays to take a few extra minutes and weigh your powder charge here and there.

One note Progressive press owners should consult the manual for their press before removing and reinserting a hull from both the powder and shot charging stations. Some presses have auto shut-offs for empty shot and powder stations, meaning that, if a hull isn't present, it won't drop the charge there. Other presses are less sophisticated and provide only a way to manually keep the bar from shuttling to the drop position for that station, while still others require you to stick in an empty hull (spent primer intact) to catch whatever drops, after which you would pour the contents back into its respective hopper and start the sequence again. You must pay attention to the progression at all times! Forgetting where you left off when removing and reinserting hulls at various points in the reloading process will leave you with a shell that has no powder, one that has powder and shot but no wad, and other disastrous issues, (such as powder and shot spilling all over your bench and jamming up your press). None of these are good for your shooting or your gun. If in doubt at either any stage, remove all hulls in all stages, tip your charge bar back and remove the hoppers, and cycle the press until it reaches the final rotation where the charging bar no longer shuttles to a position to drop anything (such as will happen first after the powder drop, and second at the shot drop on the very last shell moving through your press rotation). It's always a better idea to start over clean than wonder about that handful of "finished" shells sitting on your bench.

4. After your hull has the received the proper measure of powder, inserting a wad over the powder charge is the next step. Modern wads are available with specific heights and thickness and, as it is with other crucial components, one wad does not fit all. There are, literally, dozens upon dozens of different wad types for nearly as many different applications. Now, that being said, there is some leeway with substitution. Aftermarket wad makers like Claybuster manufacture wads that are duplicates of the original factory design (or at least close enough). So, if you can't find a Winchester 1-ounce AA wad, you can absolutely substitute the Claybuster wad *that specifically says it's a substitute for that particular wad.* What you cannot do is substitute a Remington Figure 8-type, or a Federal, or a Fiocchi, or anything else that doesn't explicitly state it's a direct substitution for the 1-ounce Winchester AA wad.

Wads need to be seated correctly on top of the powder and there also needs to be a certain amount of flex at the top of the wad in order to permit proper crimping. To seat a wad, set the wad as straight as possible in the mouth of the wad guide or directly up onto the seating sleeve, which will be above the waiting hull. Lower the arm of the press, and the wad seating sleeve/tube should press the wad firmly and evenly through the guide and into the hull. Do not apply so much force that you might tear, crush, or otherwise distort the wad. Of course, unless you were using a clear shell, you won't see the wad seating operation in action, so you must develop the feel for it, not too light and not too heavy. Simply bring the lever of the press down in an even and steady motion. You'll feel the resistance as the seating tube pushes down

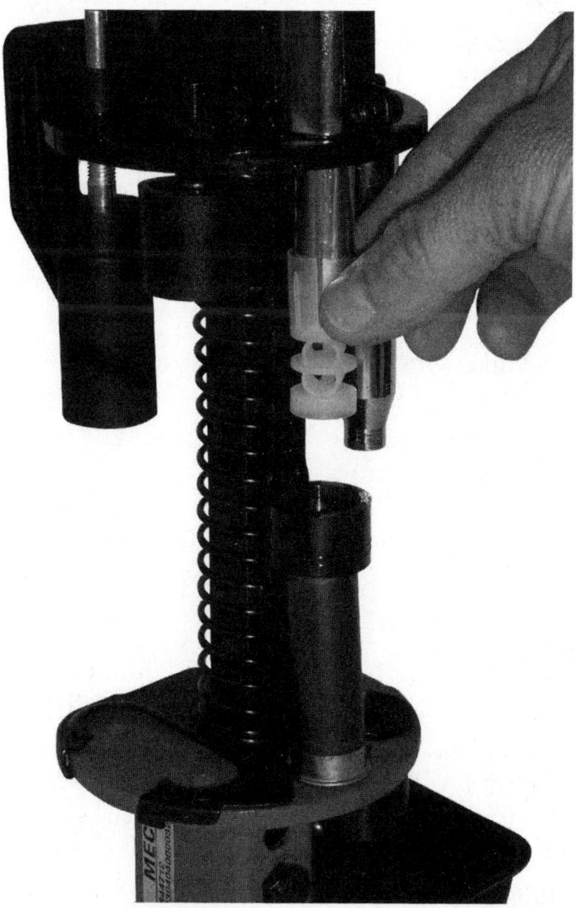

Placing the wad in position for insertion over the powder. It can only go in one way correctly or you will be picking shot up off the floor for an hour.

on the wad, and you should feel the natural, firm stop at the bottom of the lever stroke. Don't go further than that and mash the wad against the powder. Not only can this cause ignition and pressure issues, but crushing a wad causes it to cant or lean inside the hull. This effectively destroys the midsection and ruptures the gas seal, and your load becomes unreliable.

While you'll eventually develop a feel for when the wad is seated correctly, a workable method of wad seating is to observe the top of the wad as it relates to the hull's crimp folds. The top of the wad petals should sit just below the crimping line. In most, but not all, instances, this will be the proper wad height for your shot drop. If minor alterations up or down are necessary, you can do so by adjusting the pressure you apply (hydraulically or by hand) to the press ram. Some presses, such as MEC's, also have a wad pressure gauge, which coincides with wad pressure recommendations published by some wad manufacturers.

If, for any reason, the wad protrudes from the top of the hull or it drops completely out of sight inside the hull, something is wrong. Recheck your measurements, including verifying the load data, and try again.

If your press uses wad guide fingers, get used to the fact that they occasionally must be replaced. These thin, usually plastic, fingers extend into the mouth of the hull, easing passage of the wad. Modern high-performance wads are designed for a precise gas seals, and they require careful handling at this stage. Tearing or crimping the gas seal, as I've said, compromises the performance of your load.

One last note on wad seating. Hunting loads often require that a filler wad be placed inside the wad/shotcup at this point. A filler wad can be felt, cardboard, or cork. These wads are shaped like a disk and come in varying thickness. Should your load require a filler wad (or two), you may find it's easier to load these with a single-stage press. In doing so, you'll place the filler wad into the wad guide, just as you would like it to sit inside the wad column. Then use the wad rammer to gently seat it inside and on top of the shotcup's base *before* the shot is dropped.

5. Time to drop your shot charge. Shot drops and meters through a charge bar or a replaceable or adjustable bushing. Loose shot fills a cavity inside the bushing, which is cut to the specific diameter required to accommodate a certain payload. Although it seems a little primitive in the digital age, this method works quite well, especially with lead pellets No. 4 and smaller. For common trap, skeet, and sporting clays loads, you simply install the proper bar or bushing, say, a bar labeled 1¹/₈-ounce, into your press and proceed (shot sizes No. 7½ and smaller flow like water and measure almost perfectly). As shot size becomes larger, pulling consistent loads can become a problem. Larger pellets occupy space, just like smaller pellets, but those larger pellets need more "elbow room." In any given space, there is more air around and between larger pellets than there is around smaller sizes. You will also notice a slight weight reduction in the full charge, regardless the charge bar or bushing, for every shot size increase over No. 6. So, when using any particular bushing, always verify the shot weight by pulling a sample or two after first settling the shot in the reservoir. Then cycle the machine normally several times. This only takes a minute.

Sometimes your shot load will prevent crimping, but you already know the rule: Anything that affects proper crimping needs to be changed right away. If improper crimps are discovered to be related to your shot charge payload, something will have to give. Use your scale to weigh and verify shot payloads and adjust as necessary.

Unless specifically marked, shot bushings are normally intended for metering lead shot only, and that indirectly. The volumetric measurement of a bushing is a relative measure of a specific weight. That volume represents an approximation of a desired weight of shot. Perhaps this system isn't flawless, but it is relatively efficient. (To measure every payload by actual weight quickly becomes cumbersome, unless you are only loading a few hunting shells.) Steel and non-toxic shot vary from the standard of lead. Specific bushings or settings for adjustable, rather than replaceable, bushings have been worked out for steel, bismuth, tungsten-polymer, and other non-toxic shot pellets. These settings cannot be interchanged with those for lead shot.

6. Crimping your shells takes up the final stages in developing a complete load. A perfect crimp is necessary to develop the load's exact pressure ceiling. Too deep and pressure builds unnecessarily. Too shallow and the shell can literally fall apart, allowing the shot to dribble out of the hull.

Developing the crimp and starting and finishing its seal are normally a multi-station process. For different payloads, powders, and other interior components, you will need to carefully check and perhaps adjust the crimp stations of your press. Different presses also feature varying numbers of crimping stages. Experience here probably requires that a few shells in each new load will not be perfect before you decide what precisely is required. Crimping is not a static setup—you can't think of it as set it and leave it.

Unless they have a roll crimp, your hulls will have either a six- or an eight-point fold. Traditionally, heavy hunting loads filled with larger size

A good crimp is extremely important to successful reloading, because it ensures that the correct compaction of the load in the shell will create the right amount of compression. Once you pull the trigger, it is this compression that causes pressure to blow your shot out the barrel without blowing out the chamber or causing a dud round.

shot used the six-point crimp, while the relatively lighter target loads used an eight-point crimp. Nevertheless, there is no reason you have to continue this practice. A ballistics laboratory could not quantify a difference in efficiency between the two different crimp styles. So, while you can load a hunting type recipe in an eight-fold crimp hull and a target load in a six-fold crimp, what you cannot do is turn an eight-point into a six-point or vice versa.

To determine which crimp starter to use in your press, simply count the folds in the top of the hulls you're reloading. You can do the same with the crimp-starter itself if the six-pointer and the eight-pointer look alike on the outside, as they often do. It is important to use the proper crimp-starter, because folding a six-point hull with an eight-point crimp starter is going to make a mess of your hull.

Most shotshell reloaders choose to recycle previously fired hulls. Still, some like to work with new hulls. If that's your choice, follow the hull manufacturer's advice for crimping type. Also know that new hulls may be skived or un-skived. Skived hulls are thinned and taper to a fine edge at the top rim, whereas un-skived hulls have thicker sides and are untapered. At the top, an un-skived hull seems as thick and as bend-resistant as a dime. This thickness matters, because a thinner hull, as in a new skived hull, can be folded in almost any manner and, thus, works well with either a six- or eight-point crimp. The thicker un-skived hull works best with a lesser amount of folding and should usually be reserved for the six-point crimp.

Sometimes, the stacking of large shot makes crimping of any kind a chore. The large pellets will occasionally bulge upward, making hull closure difficult, even off-center. The size of the shot is just one of those variables, like the kind of hulls you use and the machine you have, that may force you to experiment a bit with closing your shells. Simply put, some machines just work better with six-point crimps, others with an eight-point.

Closing your shell is usually a multi-stage process. After the shot charge has been dropped, the first stage of this is the starter crimp stage, doing just as it says, starting to press the folds downward at their seams; if you overdo it with this adjustment, the crimp may smash together in the center during the final stage. Now, keep in mind that, in applying crimps, we often refer to a hull's "memory." This means that once plastic is creased, it will usually return to this shape the next time you bend it at those creases. New, unfired hulls, don't have that memory, of course, so you may have to have their first crimp's folds introduced

These shells illustrate improper and proper filling of the hull. The hull on the left has a very deep, concave crimp, while the hull in the center appears to be overfilled. Incorrect hull fill and crimp closure will affect the pressure inside the chamber and barrel of your shotgun and will negatively affect shell performance. The hull on the right has the proper crimp.

to the plastic by slowly working the hull into the crimp starter station a couple times, before moving on to the final crimp station. This is especially true of hulls with thick, un-skived tubes. For such hulls, when you can plainly see that folds have been introduced to the hull tube, you can proceed to the final crimp station.

7. Whether it's the next stage or two stages later, the final crimp station closes the hull, leaving what should be a flat, level surface across the top. You want the center hole where the tips of the folds meet to be as small as possible without being crushed together completely, and they should form a spiral in appearance. Shot must not dribble out of this hole! Likewise, you don't want a crimp that's so deep that the top of the shot cup is pinched beneath the edges of the crimp. If this happens, the stout plastic shot cups used in heavy hunting loads, for instance, can lift the hull plastic right out of its brass rim when you fire. Failing that, it may stretch or tear the hull in half and cause your chamber pressure to climb dangerously. (Such a situation can be remedied by adding a filler wad beneath the shot, because this raises it slightly. Many hunting loads can also work better when you top them off with a thin, .030-inch overshot card wad, which you can cut from a Manila folder, if you wish.)

In general, the folded crimp will provide you with a positive closure that is about $^1/_{16}$- to $^1/_{10}$-inch deep. Overly deep crimp centers contribute to higher pressures in loads. However, all hulls are not created equal, and proper crimping depth requires some applied judgment on your part. Work towards a depth that is sufficient to hold the crimps together. Use the crimps you see in factory ammo as your guide to your reloading crimp quality. These days, the crimp depth, quality, and finish of factory ammunition are outstanding, and every reloading tool on the market today can be adjusted to produce just as good a crimp as those from the factory.

Factory loads look good because manufacturers are automated to perfectly produce millions of lawsuit-free and performance-enhancing shotshells. These manufacturers use a formula of perfectly sized new components and an exact amount of shot with the component column height set precisely to produce a beautiful working crimp. In reloading, of course, you're not working with such perfect processes or components.

Crimping problems most often occur when components are mismatched. For instance, a low-volume, compression-formed hull is not conducive to loading heavy magnum shot and slow burning powders. These mismatched loads can become difficult to work with and are often inconsistent in the crimping stages. Too, this is yet another instance where component swapping would cause issues. Beware of component

Your finished reloads will be a source of pride—and when you fire it successfully to break a clay or bring down a quail, your second feeling may be relief!

swapping, too, because the difference in crimp height between hulls can affect your load performance. A change in hull type can undo an otherwise good fitting load, perhaps offering a too deep or a too shallow crimp. Modifications to the reloading tool will not sufficiently provide for the 1/10-inch (even more in some instances) hull depth differentials found between low-base hulls and high-base hulls, and many hulls, even from the same manufacturer, have quite different internal depths. It continues to bear repeating: do not swap components, no matter how similar they may appear.

An alternative to the six- and eight-point crimping is roll crimping. This crimping style offers a number of benefits over the more conventional folds. It's easy to obtain excellent roll crimps for every shotgun gauge, but you cannot alternate between fold crimps and roll crimps on the same hull. This would be the equivalent of component substitution, which is against the essential commandments of safe and effective reloading. Of the two types of crimps, roll crimps actually demonstrate greater load-to-load consistency of tested speeds and pressures.

Because you use an over-shot card, the length of the hull area required to form a roll crimp is about half the length required for a fold crimp. Therefore, much more of the total overall hull volume is available in a hull utilizing a roll crimp. As a side note, most factory slug loads

Three Crimping Incidentals

1. Filler Wads—When using filler wads to raise the height of the shot column for a solid crimp, be aware that the size of the shot used can influence the height of the column inside the wad. A 1¼-ounce load of No. 4 sits lower in the shot cup than 1¼ ounces of No. T. Filler wads in the base of the shot cup are there for the convenience of the reloader. Use them only if necessary and use the correct height filler wad when you do.

2. Steel Loads—For steel loads, fold crimps must be seated tightly on top of the shot column. Worn-out hulls or those made with weak, laminated plastic are not good for loading steel shot. Compression is lost when laminated hulls fall apart, thus wasting your loading efforts and components.

3. Burn Rate—Slow burn rate powders (used in most hunting loads) and weak crimps do not mix. Furthermore, in cold weather, where many hunting loads are used, these factors come together to cause dud rounds. Efficient and consistent ignition requires attention to a lot of factors, including your crimp.

4. Firm Crimps—This situation can often be remedied by adding a filler wad beneath the shot, because this raises it slightly. Many hunting loads by the way are happy when you top them off with a thin, .030-inch overshot card wad, which you can cut from a Manila folder if you wish.

use roll crimps to secure their slugs, both with and without cards. Besides superior seating of the projectile, a benefit of roll crimped slug loads is load identification. The shooter can see at a glance whether or not the load in hand is a slug load.

8. As a final step in your reloading process, you will need to mark your newly made ammunition. Since reloaded shells may have nothing at all in common with the original markings or their original boxes, you need a way to identify what it is you actually loaded. (This will save you from having to cut one apart in the field to separate your duck and goose loads from your trap loads!) A permanent marking pen will do the job, as will special shotshell stamps from Midway and Ballistic Products. There are also clear plastic shotshell boxes available from makers like MTM Case-Gard, in which you can neatly place your reloaded shells in and then note the load data with a permanent marker directly onto the boxes themselves.

19 Reloading Buckshot and Slugs

The majority of shotshell reloaders are clay bird shooters. Of course, that means the majority of reloading products out there is geared toward target shooters, and that's what you'll also find the most information on. Of course, shotgunners of all types participate in the hobby of reloading, so let's take a look at the two smaller sectors of shotshell reloading, loading for buckshot and slugs.

BUCKSHOT

Shooting buckshot means playing with big pellets, lead balls larger than BB-size, which is .180-inch. Although the most talked about load in fiction may be double-ought buck (marked on factory boxes and by gun writers as "00"), practical buckshot sizes range up to No. 0000 (.380-inch) for a modern shotgun. That's a huge ball of lead shot!

The intent of hunting with and reloading buckshot is to drive several good-size, high-energy balls deep into a relatively close-range target. Though mostly intended for big-game hunting, buckshot has also proven its effectiveness in jungle warfare and other personal-defense situations.

Buckshot sizes, left to right: No. 0000 (0.38-inch, 85 grains), No. 000 (0.36-inch, 70 grains), No. 00 (0.34-inch, 54 grains), No. 0 (0.32-inch, 48 grains), No. 1 (0.30-inch, 40 grains), No. 2 (0.27-inch, 29 grains), and No. 3 (0.26-inch, 23.5 grains). Not pictured are No. 4 (0.24-inch, 20.3 grains), No. F (0.22-inch, 16.1 grains), and No. T (0.20-inch, 12.8 grains). At 0.17-inches in diameter, No. B is occasionally found in buckshot loads.

I am often asked if finer lead birdshot loads are convertible to buckshot loads? Yes, sometimes they are. Many tests indicate that buckshot in a load comparable in weight to that of fine lead birdshot produces the same or even slight less pressure. Actually, buckshot may be used in almost any load recipe, if the weight of the shot payload doesn't change and the shot cup has room for the larger sized pellets. (Do not, under any circumstance, presume you can make a heavier-weight shot charge load than is specified in your recipe. If the recipe is for a 1¹/₈-ounce load, it's not a 1½-ounce load, even if you could fit that amount of shot in the hull.)

One of the beauties and difficulties with buckshot is that so few actual pellets are involved (although, at this size, they may best be considered balls, rather than pellets). Hard, uniform lead spheres produce the highest quality buckshot loads. At one time, you could reload copper-plated buckshot. That item is hard to find these days, although nickel-plated shot is available. Furthermore, to ensure hardness, you will want a pellet with the highest antimony content, one certainly no less than three percent.

It is important that the buckshot you use is consistent in size; take your caliper and measure a selection of individual pellets from a lot you've purchased and verify that they correspond to baseline data before using them. If you come up with different weights or measurements, weigh each of your loads, staying as close as possible to the total payload weight guidelines. Because these are hunting loads and you will shoot relatively few of them compared to what you'd shoot even casually in trap or skeet, weighing each load is not an undue burden on your time and effort.

LOADING FOR BUCKSHOT

Even though you may lack some particular component or you do not wish to purchase an additional component that you may use only for a couple shots all season, buckshot loads are assembled and tested in a specific configuration and this configuration must not be changed or modified.

INSIDE THE SHELL

Two things to remember, before we get to the details. With compression-formed hulls, recall that the hull walls become thicker toward the integral base wad and that the base is generally more bowl-shaped than flat; with straight-walled hulls, the base wad is a separate part. The tapered wall of compression-formed hulls creates a unique condition. It is wonderful for lighter loads, but, due to the size of buckshot pel-

Alliant's Blue Dot powder is a frequent component of buckshot loads and one you should find ample loading data for. A good powder to keep on hand if you're going to load buckshot with any kind of frequency.

lets, there's a need to create a condition called "stacking." Stacking is the careful placement of large pellets to best utilize available pellet payload space. Manufacturers of compression-formed hulls, such as Winchester, do not utilize this particular hull design when making large-size buckshot pellet loads. Clear Fiocchi hulls, on the other hand, make excellent containers for buckshot reloads, as they are straight-walled and have a flat, low, plastic disc base. What this means is that, generally, you will want to stay with the straight-walled hulls for the best fit and finish with large payloads, but whatever hull you select can be made into a fine buckshot load, as long as that hull's design is considered.

Some factory wads are suitable for loading buckshot, but some negatively affect an otherwise performance-balanced load. The handloader needs to remember that the buckshot size selected for a particular load has a great influence on component options. Some loads are designed in a way to accommodate smaller buckshot, but, as pellets increase in diameter, it becomes increasingly difficult to maintain the weight and bulk of the load design. Buckshot fit—ball diameter and stacking relative to gauge and components—can either make a load or hinder an otherwise sound load combination.

(Photo courtesy Ballistic Products)

TEFLON-WRAPPED BUCKSHOT

Created by DuPont scientists in the late 1930s, Teflon is a wonderful plastics product that has applications ranging from slick, non-stick frying pan surfaces to

In addition to its use on buckshot Teflon wrap can be rolled into a cylindrical shape around this slug and inserted into the hull. The slick sheathing allows easier slug passage through the barrel and choke. Use a gas seal on top of the powder plus any cork or fiber wads needed to ensure a good, tight crimp.

friction-reducing lubricant sprays. As a thin film, Teflon is an ideal substance with which to encase lead shot and slugs for a ballistic advantage.

Relative to barrel diameter, buckshot is quite a bit more imposing than something like No. 7½ birdshot. Soft lead buckshot pellets not only stack poorly in the small space of a cylindrical hull, once expelled, they have a tendency to want to stick inside a steel barrel and clog inside the necked-down choke. In contrast, buckshot with a Teflon wrap allows easier shot transition through the barrel and its constricting choke area; easing choke passage allows larger buckshot pellets to retain their shape for better shot patterns and concentration on target. Teflon wrap also allows the reloader to better utilize a wider number of different buckshot pellet sizes in hulls that are otherwise restrictive in this capacity.

Some would consider including a Teflon wrap in your buckshot reloads to be an advanced reloading technique, but really it's not hard to do, and supplies are easily available from companies like Ballistic Products, Inc. (see the resource listing at the end of this book for contact information). Teflon-wrapped loads require that a gas seal be placed on top of the powder charge, followed by an appropriate number of filler wads to form a column of the needed length. You may use either felt or cork wads of the same gauge as the load. First, however, place the coiled Teflon wrap into the hull on top of a plastic gas seal and allow it to uncoil inside the hull. Then, place your filler wads inside the Teflon wrapper and tamp them to the bottom. Stack the buckshot according to directions in your loading data, place an overshot card wad on top of the shot column, and then crimp.

PELLET STACKING

In real-world applications, buckshot stacking can either be an option or a requirement. For instance, larger pellets require stacking into some semblance of order, in to prevent creating height and crimp closure problems. Available internal space for a particular payload will make the determination for you.

The principle behind stacking buckshot pellets is that the greater the angle of surface contact between two pellets, the greater the tendency those pellets will impart energy in that direction. In other words, any deliberate order given to larger buckshot inside the hull is going to reduce the amount of apparent randomness of the shots' flight. Conversely, with smaller buckshot pellets, the quantity of pellets makes up for pattern density and reduces that same randomness; the compactness and increased surface contact of the greater number of pellets reduces the need for stacking or layering, as influences remain more consistent on all sides.

Stacking pellets in a load may require a wooden dowel to tamp the pellets into place. A slight touch of force is fine, because many buckshot loads need firm pressing to seat them properly. A layer-by-layer approach is often recommended, but only so many buckshot pellets of a certain size can be placed in a limited cylindrical volume, regardless the pressure applied from the top to compress the load stack. Stacked pellet layers are usually arranged with the alternating layers set in the valleys between the pellets of the layer below.

TYPES OF BUCKSHOT

The 85-grain pellet known as "quad-buck," or No. 0000, weighs just a few grains short of $^1/_5$-ounce. In the smaller gauges, these .380-inch pellets are difficult to work with, but they ordinarily load just fine in 10- and 12-gauge hulls. Certain loads can be assembled in the sub-gauges, though, and the quad-buck can be used as a slug in the .410-bore. This buckshot ball is excellent for big game at close range.

Stepping down slightly in size, the .36-inch triple-ought buck, or No. 000, weighs 70 grains per pellet, or $^1/_{100}$-pound. In 10-gauge hulls with an inner Teflon wrap, you can stack triple-ought in layers of three. Twelve of these large pellets create a powerful two-ounce load that's appropriate for bear (where legal, of course). The triple-ought in the 12-gauge lines up very well in layers of two. Since the triple-ought pellet does not require many layers before it exceeds the weight that can be reasonably shot in your 12, this buckshot size has often been overlooked in 12-gauge loads. Again, a slick Teflon-wrap works well as a shot carrier. This is an excellent size for big game such as deer and antelope.

At $^1/_8$-ounce (54 grains) per pellet, No. 00, or double-ought, buck is the most frequently used buckshot size for security and tactical loads. Its popularity began when the U.S. Post Office decided to use loads with this size buckshot in an effort to protect its express mail cars from train robbers in the 1880s. It quickly became the standard for protective loadings. Fired from 10-gauge double-barrel shotguns at close quarters, double-ought buck was a true stopper and, in the 12-gauge, it is still the standard for many government ammunition contracts.

Double-ought buck is on the large end of what can reasonably be loaded in the 12-gauge using standard shot cups and layers of two pellets; with modern plastic shot cups, three-pellet layers cause the hull to dimple outward. This, of course, does nothing to help the looks or fit of

Single-ought buckshot and the shell ready to load, alongside a penny for size comparison. This size shot may be difficult to arrange in a hull and often benefits from buffering, but consult your recipe before you alter beyond just pellets.

(Photo courtesy Ballistic Products)

the load. Only Teflon wrap allows an improved, near-perfect fit for layers of three No. 00 buck in a 12-gauge.

The 10-gauge offers a tight layering of four No. 00 buckshot using Teflon wrap (Teflon helps prevent hull goosebumps). Five layers of four pellets results in a mighty 20-pellet load, while six layers of three No. 00 pellets will fit into a 10-gauge hunting wad to provide 18 heavy-duty long-range pellets at a total weight of nearly 2¼ ounces. Four layers of four pellets give you a 2-ounce/16-pellet load, whose impact is truly powerful. When well seated, Remington's SP10 wad will contain four layers of three No. 00 pellets for a total of 12 pellets at 1½ ounces. And, at the far other end of the spectrum, for a special .410-bore application, a thin shotcup holds two No. 00 pellets neatly stacked one on top of each other.

Many writers believe the single-ought, or No. 0, buckshot is too often overlooked in loading equations. At .32-inch in diameter and weighing 48 grains, it is actually a marvel of fit and striking power.

The 10-gauge allows a perfect fit of layers of four No. 0 buck pellets with a Teflon wrap. If the load is intended for longer range, stacking pellets is not done. The BPD10-Tuff hunting wad (from Ballistic Products), with No. 0 buck produces an effective, hard-hitting, long-range load.

In the 12-gauge, No. 0 buck stacks into three layers of three and almost begs for Teflon wraps, resulting in an excellent one-ounce load that is both low on recoil and hits hard on impact. Using Teflon wraps, a handloader can position four layers of three pellets each to build a $1\frac{3}{8}$-ounce load. Heavier loads may be made using Teflon wrap in combinations with smaller-based wads. The 12-gauge plastic shot cup offers a tight, but not unreasonable fit, for this size buckshot. Tight fit may show up as slightly raised goosebumps on the outside of the hull.

Another often-overlooked mid-size buckshot pellet is the No. 1. It is .30-inch in diameter and weighs 40 grains. The No. 1 is large enough for deep penetration, and even a lighter load will produce multiple hits, as long as you stay within the shell's effective range.

In the 10-gauge hull, the No. 1 buck pellet size will not permit tight packing; a loose or mixed arrangement is best suited. The Remington SP10 is a thin-walled wad that offers a relatively good fit for No. 1 buck when pellets are stacked in four layers of four (about 1½ ounces).

The 12-gauge and the No. 1 buckshot pellet offer a distinctive fit for large loads in rows of three pellets with a Teflon wrap. This size also offers a good fit when stacked in many 12-gauge plastic wads, as well as when stacked in layers of four and buffered in a hull with no shotcup.

How about the slightly smaller No. 2 buck with a .27-inch size weighing 29 grains? The 10-gauge offers loose confederations of layers; a double Teflon wrap may work most efficiently in some loads. This size pellet in the Remington SP10 wad will not quite stack itself in concentric rings. A little better fit is available with the 12-gauge. Using Teflon wrap, the pellets adjust themselves into six smooth layers of four pellets.

At .26-inch in diameter and weighing 23.5 grains, a No. 3 buck pellet is small enough for random placement into any shotcup or wrap in the 10- or 12-gauge. A 20-gauge wad holds three per layer of four layers (practically ¾-ounce), for an effective load. The No. 3 buck size runs 18.5 pellets to the ounce and is capable of dense patters at maximum range.

The final buckshot sizes are the No. 4 (.24-inch and 20.5 grains), No. F (.22-inch and 16.1 grains) and No. T (.20-inch and 12.8 grains). From No. 4 on down, there is little need to stack, as pellets automatically arrange in 10-, 12-, and 16-gauge hunting shells. With the No. F, Remington's 28-gauge shotcup will hold layers of three in four stacks for an effective 12-pellet load, but attempting to stack other loads is futile and unnecessary. No. T stacks four pellets in five rows for a total of 20 pellets in Remington's 28-gauge shot cup.

When you are shooting buckshot, carefully consider pellet size. Many authorities believe American shooters consistently use buckshot that is too large for the required shooting task. Handloaders must remind themselves that lead buckshot pellets carry a lot of energy to the target. While a bigger shot size makes a bigger hole in the intended target, often more pellets on the target get the job done just as well and, at the same time, can be less punishing on the recoil end of things.

SLUGS AND SABOTS

A shotgun slug is an extraordinarily effective, accurate, and lethal single-projectile shotgun load. To completely understand slug effectiveness, a shotgunner must first know about certain characteristics of single-projectile ballistics and transferred energy.

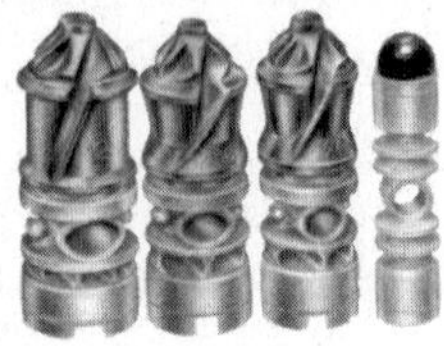

(Photo courtesy Ballistic Products)

Dangerous game slugs, from left to right in 12-, 16-, and 20-gauge and the .410-bore, perform smoothly in virtually any shotgun/barrel combination, even smoothbores. Column height is self-adjusting for perfect crimps. The Light Game Slug (right) is ideal for the 410 and 28-gauge.

There are specific differences between shotguns and rifles, of course, and one of them is velocity. A .30-06 rifle pushes a 180-grain bullet somewhere around 2,700 feet per second (fps). Even though the bullet is relatively light compared to a shotgun slug, it transfers a tremendous amount of energy to its target. Immediately upon impact, the bullet, by design, begins to deform. This deformation, or "mushrooming," is akin to slamming on the brakes of your truck. Unless there is a complete pass-through, the target absorbs all the energy carried by the projectile; this, too, is by design. Slugs work the same way, except that weight substitutes for velocity. Instead of 180 grains, we shoot 450 grains, well more than twice the weight of the rifle bullet, and instead of 2,700 fps, the slug moves at 1,400 fps, about half the bullet's speed. Yet, in a heads-up comparison at an advantageous range for the shotgun, both weapons deliver similar energies.

A reloader can change the ballistic characteristics of slug loads and gain a couple hundred feet per second. The downside is that the slug loses that extra energy quickly. Some loads, while performing with extreme accuracy on paper, just do not work well in the field. Always remember that even if some sabot manufacturers talk about tack-driving shots at 300 yards, long-range shooting is not the forté of a shotgun. Though some shotguns are capable of more, most are most effective under 100 yards (some argue for half that distance, to keep patterns tight with raw slugs).

SHOOTING SLUGS

A heavy slug is about as dainty as a charging rhino. There are a lot of options for your own charging rhino today besides the standard Foster-style slug with a round nose and concave base. (Improved Fosters for both smoothbore and rifled shotguns are still available for the 12-gauge in .660-inch diameter and one-ounce sizes.) Very light slugs can also deliver extreme accuracy and, at 100 to 200 yards, can rival the accuracy of hunting rifles. With these smaller slugs, there is a trade-off in the ballistic equation. One might really drive tacks at 100 yards, but, if it blows through a deer without bringing it down quickly, you would have gotten nowhere and failed at making a good and ethical shotgun kill. Consider your target and shooting environment before making a decision.

Light game slugs can be purchased for the 28-gauge and .410-bore. The 28-gauge is .505-inch in diameter and weighs 183 grains. The .410-bore slug is .375-inch and weighs 93 grains. In the 12-gauge, a light slug weighs 7/8-ounce or less. Heavy 12-gauge slugs run 1 1/8 ounces and sometimes up to 1¼ ounces. Most of the effective designs for the 12-gauge fall somewhere between the two weights.

A heavier slug, by virtue of its weight, carries additional energy into the target, and killing energy is achieved through a combination of velocity and weight. Any combination of an increase in velocity and weight of a slug will certainly increase its energy payload. Considered in this manner, at normal slug ranges of 50 to 100 yards, where the slug's momentum is still sufficient to resist the force of gravity, a shooter can take advantage of heavy and not-so-aerodynamic designs.

The diameter of a slug also works to the advantage of increased energy transfer. Just as weight and velocity are factors of lethality, the large diameter of a shotgun's bore allows that much more impact area. Your goal is to use as much of that bore diameter as possible, while keeping your slug stable in flight.

Ballistic data points out that a good slug load must have an internal balance for optimum performance. Too much of a good thing (i.e., weight), reduces launch speed and, therefore, produces a dramatically arced trajectory. The heaviest slugs also require slow burn-rate powders, but, remember that these are unsuitable for consistent loads in cold weather, and that's where many hunters find themselves using slugs.

OPTIMIZING SLUG WEIGHT AND SHAPE

Shotgun gauge is determined by how many round balls can be cast from one pound of pure lead in that particular diameter. Thus, one pound of lead cast into 12 round balls produces the diameter of the 12-gauge, .729-inch around and weighing 583.33 grains. That is a bit over 1¼ ounces.

In any given gauge, a round lead ball that fits the bore is about the prime computed overall weight for optimum speed and accuracy for that particular barrel. Interestingly, using this formula produces a slug that is also least susceptible to tumbling and takes full advantage of the barrel's diameter to offer the broadest possible face to the target. (Incidentally, round balls are not a ballistic panacea either, but certainly are a reasonable and effective solution to many slug-shooting needs. However, you should never fire a bore-sized ball from a choked barrel, because the ball will collide with the choke and both may very well be deformed as a result.)

It is important to keep an eye on the volume a projectile occupies inside the barrel and not wholly fixate upon the projectile's weight. Shotgun gauge/barrel optimums have more to do with volumetric displacement than you would expect. Since slugs are not usually made of pure lead but are, instead, often an alloy of metals, the volume displacement is greater than an equal weight chunk of pure lead. Therefore, the optimum slug will be both slightly larger and lighter than an optimum-sized slug of pure lead. Carrying forward this formula we find that:

- The 10-gauge slug optimum is less than 1⅝ ounces and close to 1½ ounces.
- The 12-gauge slug optimum appears to be somewhere between one ounce and 1⅛ ounces.
- The 16-gauge slug optimum is just shy of one ounce.
- The 20-gauge slug optimum is between ¾-ounce and ⅞-ounce.
- The 28-gauge slug optimum is about a ½-ounce.

SLUG LOADS AND ROLL CRIMPS

The conventional 12-gauge hull is 2¾ inches long when it is fully open and extended. Hulls loaded with slugs are usually roll-crimped down to the top of the outer shoulder of the slug, leaving the correct overall *folded* length. If the hull is new and unfired, roll-crimping produces an effective closure. For accuracy and pressure consistency, slug handloaders should strongly consider using only new hulls for their roll-crimped loads, and certainly not recycled hulls that have previously had six- or eight-fold crimps.

SLUGS FOR RIFLED BORES VERSUS SMOOTH BORES

If your shotgun barrel is rifled, you need to take advantage of it by using one of the slugs designed to work harmoniously with it. This type of slug, by design, is more accurate at extended range. These slugs have rifling pickup rings molded directly into their bodies, rings that will "grab" the barrel's rifling. A little experimentation will lead you to the one that works best with your gun. You should shoot it at various ranges until its performance becomes predictable.

Smoothbore purists can, and sometimes do, attain high levels of accuracy with specialized slugs for their guns. To maintain their accuracy, these slugs must have proper fit within the barrel. Most often, inaccuracies associated with smoothbores and slugs can be directly attributed to a slug's deflection during its course of travel down the barrel. Deflection can sometimes be cured with a Teflon wrap around the slug. This creates the snug fit between the slug and bore needed for accuracy and also allows the slug to slip down the barrel in milliseconds without causing or encountering extreme friction. On that note, although smoothbore barrels will not impart any negative influence upon shotgun slugs, neither will they impart any positive influence.

OTHER SLUG LOADING CONSIDERATIONS

With the modern slug's superior design and balanced weight, the handloader can produce high-speed, personalized loads for the popular, longer 3-inch chamber. If your shotgun is so chambered, you may wish to use the additional shell length to build performance, as long as any change will conform to the shotgun's design and pressure intent. The longer hull allows a bit more space for powder and other components, including additional gas seals, which can help obtain superior consistency and velocity.

For best results, shooters need to take some time to consider important ballistic variables associated with shooting slugs. Two factors affecting what you want to put into a round are the expected temperatures at the time of the hunt (affecting the powder burn and velocity) and the landscape of the hunting area (long or short range).

Some slugs have a tapered or spired front end. This can be good for some particular hunting applications, but you never want a sharp point to stick out beyond the hull's top edge, especially if you are using a gas gun or a pump with their tubular magazines; it is possible for the protruding slug tip to contact the primer in the shell ahead in the magazine during rough handling and during recoil, creating a possible and catastrophic chain-fire event. If you add an overshot card to the top of the load before applying a crimp, you can avoid this problem.

Intended for rifled barrels, the BRI slug has a distinctive hourglass shape. Housed in a sabot that sheds away after it exits the barrel, these slugs are devastating on game.

The Collet Cup Sabot Slug is designed for rotation and high velocity in rifled barrels. The 12-gauge slug is a 300-grain (11/16-ounce), .45-caliber jacketed, hollowpoint bullet wrapped by a molded plastic sabot.

The Brush Buster Myth

Except in the South, where many hunters shoot over large, wide-open food plots, most favorite hunting areas east of the Mississippi River are filled with dense cover. In order to hit your target, you may have to fire through a maze of branches, twigs, and leaves. Yet, no matter what you've heard or read, there is no such thing as a "brush buster" slug or bullet. So let us not propagate this myth. If a projectile of any type ever hits its intended target after traveling through leaves or grass or anything other than air, that indeed is an accident. The only thing we can say for sure is that tougher projectiles, which are less apt to deform in such encounters, will be less deflected.

What powders are best for slug shooting? Medium to fast burn-rate powders have worked well with most types of slugs and conditions. For long-range shooting in cold weather, try a single-base powder (no nitro!) such as the Accurate Solo 1250. In colder climates, you can also help your powder burn completely by crimping the shell deeply. Slug wad columns do not require wad pressure for seating, but always need firm crimps to create conditions for proper combustion.

Finally, for an accurate slug load, a plastic gas seal should be used, sometimes more than one. Plastic gas seals allow the load to generate maximum energy from the powder burn and place the pressure directly behind the slug, keeping disruptions to a minimum.

20 Reloading Shotshells for Extreme Temperatures

Not every day in the field or on the shooting line will be a bluebird day. Maybe you're duck hunting in a freezing rain over ponds half-frozen. Or maybe you've arrived at the sporting clays course in San Antonio, the temperatures hovering in the mid-90s, with thunderstorms threatening and the humidity level at practically 100 percent. For these extremes and everything in between, the best thing about them is that, after a little bit of study and experimentation to see what works best in your gun, you can customize loads to parallel your shooting conditions. In fact, instead of going about the whole idea of reloading based on saving some money, why not approach the reloading bench with the desire to maximize your shooting performance against your shooting conditions as your driving factor? That's the kind of thinking that will change your performance, your success and, eventually, your satisfaction with reloading. Let's take a look at how to do just that.

SHOOTING IN HEAT

Hot days increase the pressure inside loads, especially in large hunting loads packed with slow-burning powders. The type of powder used however, does not have as much influence on increased pressure as does the amount or volume of powder in the load. For example, a 10-gauge charge using the largest volumes of powder, or even a heavy 3- or 3½-inch 12-gauge load, will hit you harder with recoil when you fire it in 90-plus-degree temperatures than it will in cold weather.

Here's the worst-case scenario. For a big load with a lot of powder, one can expect a rise of about 1,000 psi for every 10 degrees over 70 degrees Fahrenheit. Smaller target and upland game loads, with smaller charges of powder, have a much smaller rate of rise and a lower peak, possibly as low as 200 psi for every 10 degrees over 70 degrees Fahrenheit. (In other words, you need not make any adjustments to trap, skeet, sporting clays, or small game loads for use in high temperatures.)

In excessively high ambient temperatures, the powder's burn rate becomes the variable factor. It is not one specific powder for which this occurs, either, for all powders react to heat (and to cold), and ambient temperature must be considered if we want our shells to operate at peak performance—and by peak performance, I mean precisely as we expect them to operate, neither slower nor faster. The more powder contained in one's load, the greater the reaction to temperature one can expect.

Following are a few steps you can take to keep loads operating as they should or as tested in the laboratory at 70 degrees Fahrenheit, regardless how high the mercury. Keep in mind that while some of this (and what is also recommended in the section for shooting in the cold), may seem to fly in the face of just saying "No" to component substitution and fiddling with recipes, these are actually established practices for dealing with shotshell performance in high heat environments. If you have any questions or doubts at all, give your powder and/or primer manufacturer customer service a call, or one of the reloading experts at Brownell's, Hodgdon, and MidwayUSA, and run your proposed alterations past them. Also, if you do make any of the following alterations to your reloads, it is, therefore, incumbent on you to mark the shell so that you will remember what you have inside those hulls and for what purpose the load was intended.

1. Primers: If the load you're using calls for a magnum primer, consider switching to a standard or cooler primer. If a CCI 209M is called for, think about switching to a CCI 209. A simple primer change can affect a load by 1,000 psi and sometimes even more, a fact that should open your eyes regarding random substitutions. However, do not then use these loads indiscriminately in cool or cold weather, because the result will be shot dribbling meekly out your barrel, rather than flying with the authority you need to bring down a pheasant rooster.

Alliant says its "e^3" 12-gauge competition powder is "far less affected by hot or cold weather than single-base powders." Apparently, its burn rate and ignition characteristics are "substantially less affected by temperature extremes, to give more consistent and reproducible performance all year round."

2. Powders: If you know the day is going to be hot, think about reducing the powder charge by five percent or two grains (in the 12-gauge), whichever is greater in your particular load. This tiny change will substantially reduce pressure on a hot day, perhaps by as much as 1,000 or 1,200 psi! Do not, however, perform the reverse operation and up your powder charge to accommodate cold weather.

3. Crimps: It is an old reloader's trick to crimp hulls more deeply for cold days, as this will sustain powder ignition and create extra pressure for combustion. You will get the same result on hot days. (A deep crimp is considered greater than 1/8-inch from the rim to the surface of the crimp.) If your reloading instructions should call for a deep crimp, think about how you are going to use the loads. Some sources may not know that deep crimps are conducive to increased pressure, but you do!

SHOOTING IN COLD

Many shooters believe it is cold either from the moment when the first frost crunches under their boots and the wind causes them to wear a balaclava or only when icicles are hanging from the clubhouse eaves. For me and the performance I expect from my reloads, cold weather shooting begins when the temperature dips into the 40s and the humidity rises.

As a reloader, you can manage your loads cold weather, just as you can for hot weather. Here are a few steps you can take to keep loads operating as they should or as tested in the laboratory at 70 degrees Fahrenheit, by improving low temperature load performance:

1. Primers: It may be possible to select an alternate primer to the one specified for the recipe you are loading, one that will increase the ignition quality of a particular load. Search through established load data sources for recipes similar to your favorite, but ones that use hotter primers. CCI and others manufacture primers that burn consistently, regardless the weather. You might also explore magnum primers, although "magnum" in this case is something of a misnomer, because they are not usually specified for magnum loads, rather they explode with greater heat compared to the standard No. 209. You should probably never use these loads if the environment temperature of the shell rises above 40 degrees Fahrenheit, though, or the pressure generated could be too great for the chamber of your shotgun.

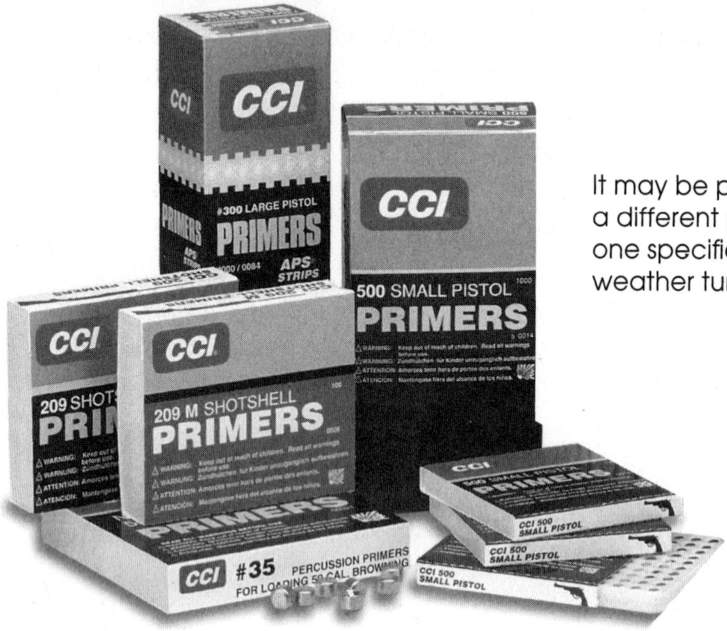

It may be possible to select a different primer than the one specified, when the weather turns bitterly cold.

2. Hulls: Hulls become brittle in frigid weather. Think of a garden hose left outside in the winter and imagine how stiff and cold it feels in your hands. In such cold conditions, restrict your reloading to hulls that a reasonable person would consider to be in excellent condition. Sure, this is true for all of your hunting loads, but it becomes especially important in cold weather. Cold exacerbates any already brittle components, and using a faulty shell in frigid conditions may cause a gas leak or blow a shabby crimp right off the top of the hull.

3. Powders: Powder is the key performance variable under any shooting condition. Various powders have very different ignition qualities and ignition is only one factor in a powder's design intent. To accommodate another area of performance, perhaps an extended burn, for example, some powders may explicitly state that they are not cold weather compatible. Also, to accommodate the medium burn-rate powders desirable for cold weather, consider using slightly lighter shot payloads.

4. Wads: Like the fuel/air mixture in an internal combustion engine, powder needs the right compression-generated pressure to burn properly. For a thorough and clean burn cycle, it needs a lot of pressure. If, however, the compression suddenly drops, your burn cycle is going to be interrupted; a lot of crud will be deposited in your barrel and your shot will perform lamely.

To substitute target wads in hunting loads is to cause your performance to suffer greatly in general and even more so in the cold. The weak pillar section of such target wads, usually designed for 1$^1/_8$-ounce or lighter shot payloads, moves forward too quickly, reducing pressure and interrupting the burn cycle. As a solution, I have heard several gunners talk about substituting high-density steel shot wads in preparation for a hunt in extremely cold weather. One apparent characteristic of the high-density plastic used to cup steel shot is that the tougher, harder product does not become a whole lot less flexible as it gets colder.

5. Shot: Count on giving away some velocity to the cold. Wind also usually compounds the dilemma of lost velocity, and this means lost pellet energy at the target and reduced effective range. Compensation begins with pellet size selection. Weight is a pellet's dominant factor in retaining momentum, so you can change this variable and actually increase a load's lethality. A single size larger is usually enough, say, from No. 4 to No. 3.

6. Steel loads have an advantage over lead in cold weather, because of the principle of bulk versus weight. This describes the relationship between the number of pellets in a given load and the pressure created by it versus the traditional weight of the pellets in a load. A large volume of pellets adds to the quick generation of combustion pressure, assisting the load in getting a better burn out of slower burn rate powders.

7. Something as simple as keeping your shells warm in the field will also help load performance. One method is to keep them inside your coat and another is to use a plastic shell box "warming house." Place a heat pack inside and shut the lid. This inexpensive act can keep your shells warm enough to keep performance consistent with that experienced during more temperate days.

Appendix I
Safety and Other Considerations

There is no question that reloading involves some dangers and requires more than a passing nod at keeping yourself safe. Nevertheless, reloading is an activity that can involve the whole family, even very young shooters, provided adequate instruction and supervision are provided. Any person who is large enough to control a gun and capable of shooting or hunting can reload ammunition successfully, but regardless the age or experience of the handloader, there are several basic rules that every reloader should follow.

- Read your loading data carefully. The biggest reloading mistake you can make is misreading or misapplying load data. In an extreme case, such a mistake can kill you or someone standing nearby when the pressures ignited in the chamber are too great for your gun's barrel and receiver.
- Never leave powder, primers, and lead shot and bullets where unsupervised children have access to it. A locked and cabinet or a safe for powder and primer storage is always an excellent idea.
- Always wear safety glasses or goggles when reloading.
- Powders contain nitroglycerin. Inhalation, ingestion, or skin contact may cause severe headaches, nausea, and a drop in blood pressure. In case of ingestion or inhalation, call a doctor immediately. Avoid contaminating food and beverages with powder residue. Wash thoroughly after handling powder and have your reloading bench setup in a well-ventilated area (but not one with fans).
- Do not smoke, use spark-producing tools, or work near an open fire or heating source such as a fireplace, lantern, or portable heater when reloading. Smokeless powder is highly flammable and primers are explosive. Powders and primers should always be stored and handled in such a way to avoid impact, sparks, flame, friction, and heat. You must have a fire extinguisher available in your reloading area.
- Pour only the amount of powder needed for your immediate work from your main powder bottle or jug.
- Clean up any spilled powders with a brush and dustpan immediately. Do not use a vacuum cleaner, which can ignite powder through sparking. Never vacuum up primers.
- Check the powder measure each time it is used. Make sure the settings have not been accidentally changed. Check the weight of thrown charges often during any given reloading session.
- Store powder only in its original container; do not repackage powder. The original manufacturer's container may look nondescript, but it was specially designed for storing powder. When you are ready to discard the carton or package, make certain that it is empty and clean; never use it for any other purpose.
- Shotshell reloaders should never use lead shot with data intended for steel shot or other non-toxic shot data, and vice-versa.
- For shotshell reloaders, know that shotshell wads differ in their sealing ability. Use only the wads specified for a particular load. Do not substitute wads that look similar; verify that a wad from an aftermarket brand such as Claybuster is manufactured as a specific substitution for a specific factory wad.

- Do not substitute smokeless powders for either actual blackpowder or blackpowder substitutes, nor vice versa. Lead and steel pellets have drastically different ballistic properties, and to mistakenly use one for the other can cause serious property damage and personal injury.
- Do not substitute primer brands or primer sizes unless your loading data makes specific reference to such substitution. All primers are not equally powerful. Some produce more gas at a higher temperature and can cause excessive and dangerous pressures. Use only the primers specified for a particular load.
- Never substitute one powder for another, and never mix types of powder, even if one manufacturer says a certain powder is "similar" to that made by a different company. As with other components, follow load recipes exactly.
- Do not substitute components, exceed listed maximums, or load less than listed minimums.
- Discharging firearms, cleaning firearms, or handling ammunition in poorly ventilated areas may result in exposure to lead, an element known to cause birth defects, reproductive harm, and other serious injuries and health issues. Have adequate ventilation at all times when reloading and shooting. Wash hands and face thoroughly after handling lead and before coming in contact with food, chewing materials, and smoking materials.
- Find a place and time where you can focus on loading. Building safe and consistent loads requires your total attention. Additionally, most companies suggest that establishing a reloading routine at the bench will result in the uniform loads and minimize the chance of loading errors.

KNOW YOUR GUN

You can reload for any shotgun, rifle, or handgun, but semi-auto actions require some special attention.

Compared to a semi-auto that bleeds off some propellant gases, using them to cycle the mechanical action, an over/under is a relatively simple mechanical instrument. You pull the trigger and everything goes out the muzzle or is ejected when the action is broken open. You can shoot practically any shell through an over/under that the chamber will hold, and you know immediately if a shell will not, for some reason, fit in the chamber, because you load each individual shell by hand. The same can be said of most single-shot and bolt-action rifles and handguns, as well as many revolvers.

A semi-auto action, on the other hand, can be much more deceiving, because, after the first shell is loaded into the chamber, subsequent shells are automatically loaded into a tubular magazine. Additionally, once a shot is fired, the expended shell is mechanically ejected (or at least it should be), and the next round fed up from the magazine and into the chamber to be fired next. The gun, in a sense, does all the thinking. The problem is that, with a semi-auto, especially semi-auto shotguns (as well as lever-action rifles and pump shotguns with tubular magazines), shells may fit in the magazine, but not in the chamber. This is especially true for shotgun hulls that have been fired more than a few times (and even though it is the first job of your reloading press to properly resize your hulls). The brass can deform slightly, and any irregularity may prevent a shell from cycling properly in such firearms. It's not something that should prevent you from reloading, but it is something to be aware of and it should inspire you to use only clean and undamaged components.

Appendix II Shotshell Reloading Powder Bushing Charts

(Courtesy Hodgdon Powders)

Lee

#	CLAYS	INTER	UNIV	HS-6	H110	TITE WAD	LIL' GUN	LGSH	TITE GRP
.095	10.7	12.3	15.2	21.9	22.5	12.4	23.5	19.1	18.4
.100	11.3	13.0	16.0	23.0	23.7	13.0	24.7	20.1	19.2
.105	11.8	13.7	16.8	24.2	24.9	13.7	26.0	21.1	20.0
.110	12.4	14.3	17.6	25.3	26.1	14.3	27.2	22.1	20.8
.116	13.0	15.1	18.5	26.7	27.5	15.1	28.7	23.3	21.7
.122	13.7	15.9	19.5	28.1	28.9	15.9	30.2	24.5	22.9
.128	14.4	16.6	20.4	29.4	30.3	16.7	31.7	25.7	24.1
.134	15.0	17.5	21.4	30.8	31.8	17.5	33.1	26.9	25.3
.141	15.8	18.4	22.5	32.4	33.4	18.4	34.9	28.3	26.5
.148	16.6	19.3	23.6	34.0	35.1	19.3	36.5	29.8	-
.151	16.7	19.6	24.0	-	-	19.7	37.2	30.4	-
.155	17.3	20.1	24.7	35.7	36.7	20.4	38.2	31.1	-
.163	18.2	21.1	25.9	37.5	38.6	21.4	40.2	32.7	-
.171	19.1	22.0	27.1	39.3	40.5	22.5	42.1	34.3	-
.180	20.1	23.1	28.5	41.4	42.7	23.6	44.4	36.1	-
.189	21.2	24.1	29.9	43.5	44.8	-	46.6	37.9	-
.198	22.1	25.2	31.3	45.5	46.9	26.0	48.8	39.7	-

700-X	800-X	PB	7625	4756	WST	WSF	WAA LITE	SUPER HCP
10.9	11.0	12.1	14.4	14.7	12.1	18.1	12.0	17.4
11.5	11.6	12.7	15.2	15.5	12.7	19.0	12.6	18.3
12.0	13.0	13.4	15.9	16.3	13.4	20.0	13.3	19.2
12.6	13.7	14.0	16.7	17.0	14.0	21.0	13.9	20.1
13.3	15.9	14.7	17.3	17.9	14.8	22.1	14.6	21.3
14.0	16.7	15.5	18.5	18.9	15.6	23.2	15.4	22.4
14.7	17.5	16.3	19.4	19.8	16.3	24.4	16.1	23.5
15.4	18.3	17.1	20.3	20.7	17.1	25.5	16.9	24.6
16.0	19.2	17.9	21.3	21.8	18.1	26.9	17.8	25.9
16.7	20.3	18.9	22.4	22.9	19.0	28.2	18.7	27.2
17.1	20.7	19.2	22.9	23.4	19.4	28.8	19.1	27.7
17.5	21.2	19.7	23.5	24.0	19.9	29.6	19.6	28.5
18.4	22.3	20.7	24.7	25.2	21.0	31.2	20.5	30.0
19.3	23.4	21.8	25.9	26.5	22.0	32.9	21.5	31.5
20.4	24.6	22.8	27.3	27.8	23.2	34.4	22.7	33.2
21.4	25.9	24.0	28.6	29.2	24.3	36.2	23.9	34.8
22.4	27.1	25.1	30.0	30.6	25.5	37.7	25.0	36.5

Mec

#	CLAYS	INTER	UNIV	HS-6	H110	TITE WAD	LIL' GUN	LGSH	TITE GRP	700-X	800-X
7							11.9				
8							12.5				
9							12.9				
10					13.5		13.4	11.0			
11					14.4			11.7			
12					15.4			12.4			
A12				15.5	16.4			13.2			
13				16.1	17.0			13.9			
14				17.9				15.4	13.8		
15			12.0	18.8				16.1	14.5		
16		10.8	12.9	19.8				17.0	15.4		
17		11.4	13.5	20.7				17.8	16.2		
18		11.8	14.0	21.6				18.7	17.1		
19		12.3	14.6	22.6				19.5	18.1		12.2
20		12.9	15.5	23.7		13.3		20.3	19.0		12.9
21		13.3	16.0	24.7		13.9		21.2	19.7		13.6
22		14.3	16.8	25.8		14.5		22.1	20.2	12.5	14.4
23		14.6	17.4	26.8		15.2		23.1	21.1	13.1	15.1
24		15.5	18.3	27.9		15.8		24.0	21.9	13.7	15.7
25		16.0	18.9	28.9		16.2		24.8	22.7	14.3	16.5

PB	7625	4756	WST	WSF	WAA LITE	SUPER HCP
	12.2	12.4				14.6
	13.0	13.1		16.2	11.6	15.4
	13.6	13.7	11.9	17.0	11.9	16.1
11.8	14.3	14.3	12.5	17.8	12.2	17.0
12.6	14.9	14.9	13.1	18.6	12.9	17.6
12.9	15.7	15.6	13.8	19.6	13.7	18.6
13.5	16.3	16.3	14.3	20.4	14.2	19.3
14.1	17.0	17.1	14.9	21.3	14.7	20.4
14.7	17.6	17.8	15.5	22.1	15.3	21.2
15.3	18.3	18.5	16.2	23.0	15.4	22.1
15.8	18.9	19.2	16.8	23.8	16.3	22.8

Mec

#	CLAYS	INTER	UNIV	HS-6	H110	TITE WAD	LIL' GUN	LGSH	TITE GRP	700-X	800-X
26		16.6	19.1	30.3		16.9		25.9	23.7	15.0	16.9
27	14.6	17.1	20.2	31.2		17.6		26.9	24.7	15.6	17.6
28	15.2	17.7	21.1	32.6		18.3		28.0	25.8	16.3	18.2
29	15.9	18.3	21.7	33.6		19.0		29.0		16.9	18.8
30	16.3	19.1	22.7	35.0		19.7		30.1		17.6	19.6
31	16.9	19.9	23.4	36.2		20.4		31.1		18.3	20.3
32	17.5	20.6	24.4	37.8		21.4		32.2		19.3	21.0
33	18.2	21.2	25.2	39.0		22.2		33.3		19.7	21.7
34	18.9	21.9	26.0	40.4		23.0		34.5		20.4	22.5
35	19.4	22.6	26.7	41.1		23.4		35.6		21.2	23.4
36	20.1	23.4	27.6	43.0		24.0		36.7		21.9	24.2
37	20.7	24.2	28.7	44.4		25.3		38.0		22.7	25.1
38	21.4	25.1	29.7	46.2				39.3			25.9
38A	22.0	26.0						40.6			26.7
39	22.8	26.9		47.6				42.0			27.5
39A								43.1			28.3
40								44.4			29.1
41								47.3			31.0
41A								48.8			
42								50.2			
42A								51.5			
43								53.0			
43A											

PB	7625	4756	WST	WSF	WAA LITE	SUPER HCP
16.5	19.8	20.1	17.6	24.9	16.9	23.8
17.1	20.5	20.8	18.2	25.7	17.6	
17.8	21.3	21.5	18.9	26.6	18.3	
18.5	22.0	22.2	19.5	27.6		
19.2	22.9	23.1	20.4	29.0		
19.9	23.7	23.9	21.0	30.0		
20.6	24.8	25.0	21.8	31.4		
21.3	25.6	25.8	22.6	32.4		
22.1	26.5	26.7	23.4			
22.8	27.3	27.5	24.2			
23.6	28.4	28.6	25.2			
24.4	29.2	29.5				
25.2	30.3	30.6				
26.0	31.2	31.6				
26.7	32.1	32.5				
27.5	33.0	33.5				
28.4	34.0	34.5				
30.1	35.8	36.4				
		37.5				
		38.4				
		39.4				
		40.4				
		41.4				

Ponsness/Warren

BU#	CLAYS	INTER	UNIV	HS-6	H110	TITE WAD	LIL' GUN	LGSH	TITE GRP
1AX							12.0		
2AX							12.7		
3AX							13.1		
1A					13.7				
2A					14.8				
3A				15.3	15.6			13.0	
A				16.8	17.5			15.0	
B				17.6				15.5	14.5
C			12.5	18.5				16.5	15.0
C1			13.0	19.6				17.0	16.0
D			13.2	20.4				17.3	16.3
D1			14.0	21.3				18.2	17.0
E			15.7	23.5				20.7	19.4
E1			16.2	24.0				21.6	20.0
E2			17.2	26.1		14.7		22.8	21.3
F			18.1	27.5		15.6		24.2	22.7
F1			18.3	28.3		15.8		24.4	22.9
F2						16.4		25.3	23.5
F6			18.9						
G			20.2	31.2		17.4		26.8	24.7

700-X	800-X	PB	7625	4756	WST	WSF	WAA LITE	SUPER HCP
			12.5	12.7				
			13.2	13.3		16.4	11.4	15.5
			13.5	13.6		16.6	11.6	15.7
		11.8	14.3	14.3	12.6	17.7	12.0	17.0
	13.3	13.2	16.0	15.9	14.1	20.1	14.0	19.0
	14.0	13.7	16.6	16.6	14.6	20.8	14.3	19.8
12.9	14.9	14.5	17.5	17.6	15.2	21.9	15.0	20.9
13.9	16.0	15.5	18.4	18.7	16.4	23.2	16.1	22.1
14.0	16.2	15.6	18.5	18.8	16.5	23.4	16.2	22.2
14.6	16.6	16.0	19.2	19.5	16.9	24.2	16.5	23.1
15.6	17.6	17.1	20.5	20.7	18.1	25.7	17.2	23.3

Ponsness/Warren

BU#	CLAYS	INTER	UNIV	HS-6	H110	TITE WAD	LIL' GUN	LGSH	TITE GRP
G1			20.6	32.7		17.7		27.1	
H	16.0	18.5	22.0	34.0		18.9		29.0	
I	16.7	19.3	22.8	34.4		19.6		30.0	
J	17.2	19.8	23.9	35.5		20.3		31.2	
J1	17.6	20.3	24.5	36.9		21.0		32.0	
K	17.8	20.6	24.8	37.1		21.3		32.5	
L	18.7	21.6	25.7	39.5		22.2		34.0	
M	19.7	22.5	26.8	41.2		23.4		35.7	
N	20.8	23.7	28.2			25.2		38.0	
O	21.0	24.0	28.5					38.5	
P	21.5	24.5						39.9	
Q	21.9	25.0						40.1	
R								41.8	
S								42.5	
T								45.4	
U									
V									
W									
X									
Y									

700-X	800-X	PB	7625	4756	WST	WSF	WAA LITE	SUPER HCP
15.9	17.8	17.4	20.8	21.0	18.5	26.0	17.3	24.0
17.1	19.0	18.7	22.2	22.4	19.7	28.0	18.8	
17.6	19.6	19.2	22.9	23.1	20.3	29.0		
18.4	20.4	20.0	23.8	24.0	20.9	30.2		
19.2	20.9	20.5	24.6	24.8	21.7	31.1		
19.5	21.2	20.8	24.9	25.1	22.0	31.8		
20.1	22.1	21.8	26.1	26.3	23.1			
21.5	23.7	23.1	27.6	27.8	24.6			
22.9	25.3	24.6	29.4	29.7				
	25.8	25.1	29.8	30.1				
	26.2	25.5	30.7	31.0				
	26.4	25.7	30.8	31.2				
	27.5	26.7	32.1	32.5				
	28.0	27.2	32.6	33.1				
	29.4	27.8	34.7	35.2				
	31.7			37.3				
	32.3			38.0				
	34.9			40.1				
				40.9				
				42.2				

Hornady/Spolar

GRS	CLAYS	INTER	UNIV	HS-6	H110	TITE WAD	LIL' GUN	LGSH	TITE GRP	700-X	800-X
11.0		333									
12.0		348					250				
13.0		363			256		259				360
13.5			330				266				
14.0		375	342		266			300			372
14.7									318		
15.0		390	354			384		309	324	408	381
16.0	429	402	366			396		318	330	420	
16.5	435	408	372			402		327	336	423	396
17.0	441	414	378	300		408		330	342	429	
17.4	447	417	384			414		336	345	435	414
18.0	456	423	390	309		420		339	351	441	420
18.5	462	429	393			426		345	357	447	426
19.0	468	435	402	318		432		348	360	450	429
19.4	474	438	405			435		354	363	459	435
20.0	483	447	411	327		441		357	369	465	441
20.6	489	453	417			450		363	375	468	447
21.0	495	459	420	336		453		366	378	471	453
21.5	501	465	423			459		369	381	474	459
22.0	507	471	429	345		462		372	387	486	462
22.5		477	432			468		378	390		468
23.0		483	438	351		474		381	396		471
23.5		489	441					384	399		474
24.0		495	447	360		486		390	405		
24.5		501	450			489		393			486
25.0		507	456	366		492		396			
26.0			468	375				405			
27.0			480	381				414			
28.0			489	387				423			525
29.0			501	393				429			
30.0				402				435			534
31.0				408				444			
32.0				414				450			
33.0				423				459			
34.0				429				465			
35.0				435				471			
36.0				441				480			
37.0				444				486			
38.0				450				492			
39.0								498			

PB	7625	4756	WST	WSF	WAA LITE	SUPER HCP
					321	
345			333		342	
360	327		348		351	
			354		354	
372	339	339	360		363	
381			372		372	
			378		378	324
396	360	360	387		390	386
	372	372	393	330	402	339
414			399	336	408	342
420	381	381	405	339	414	348
423			411	345	420	354
429	390	3*90	417	348	426	360
435	396	396	423	354	432	363
441			426	357		369
444	408	408	432	363		372
453	414	414	441	369		375
456			444	372		381
459	420	420	450	378		384
468		426	456	381		390
474		429	459	387		396
480	435	435	465	390		399
486	444	438	468	396		405
	450	441	474	399		408
	453	444	477	402		
498	456	453	483	408		
507	462	459		417		
516	471	471		423		
525	480	480		429		
	486	486		438		
	498	498		444		
		507		450		
	516			456		
				459		
		525				
		534				
		549				
		558				

Appendix III
Web Resource Listing

METALLIC CARTRIDGE PRESSES, DIES, AND TOOLS

Ammo Load Worldwide—www.ammoload.com
Bass Pro Shops—www.bassproshops.com
Brownell's—www.brownells.com
Cabela's—www.cabelas.com
Dillon Precision—www.dillonprecision.com
Forster
Hornady—www.hornady.com
Lyman Products—www.lymanproducts.com
MidwayUSA—www.midwayusa.com
Pacific Tool & Gauge—www.pacifictoolandgauge.com
RCBS—www.rcbs.com
Redding Reloading, Inc.—www.redding-reloading.com
Sinclair International—www.sinclairintl.com

METALLIC CARTRIDGE BRASS, BULLETS, PRIMERS, AND LOADING ACCESSORIES

Alpha Bravo Manufacturing—www.bulletmen.com
AmmunitionStore.com
Badman Bullets—www.badmanbullets.com
Barnes Bullets—www.barnesbullets.com
Bass Pro Shops—www.bassproshops.com
Berger Bullets—www.bergerbullets.com
Berry's Manufacturing—www.berrysmfg.com
Black Hills Shooter's Supply
Brecknell—www.brecknellscales.com
Brownell's—www.brownells.com
Buffalo Arms Co.—www.buffaloarms.com
Bullet Proof Samples—www.bulletproofsamples.com
Cabela's—www.cabelas.com
Camdex Automatic Ammunition Systems—www.camdexloader.com
CCI—www.cci-ammunition.com
Cheaper Than Dirt—www.cheaperthandirt.com
Corbin Custom Bullet Makers—www.corbins.com
Crow Shooting Supply—www.crowshootingsupply.com
Custom Brass and Bullets—www.custombrassandbullets.com
Cutting Edge Bullets—www.cuttingedgebullets.com
Diamondback Custom Bullet Fabricators—
www.diamondbackcbf.com
Federal Premium—www.federalpremium.com
Forster Products, Inc.—www.forsterproducts.com

Frankford Arsenal/Battenfeld Technologies—
www.battenfeldtechnologies.com
Friendswood Bullet Co.—www.friendswoodbullet.com
Gander Mountain—www.gandermountain.com
Graf & Sons—www.graf.com
GSI, International—www.gsiinternational.com
Huntington's Die Specialties—www.huntingtons.com
Iosso—www.iosso.com
J and J Products—www.jandjproducts.com
JLK Bullets—www.swampworks.com
Kramer Industries—www.kramerindustriesonline.com
L&R Ultrasonics—www.lrultrasonics.com
Lapua—www.lapua.com
Leadheads Bullet Co.—www.proshootpro.com
Lohman Arms—www.lohmanarms.com
Magtech—www.magtechammunition.com
Hornady—www.hornady.com
Matt's Bullets—www.mattsbullets.com
MidSouth Shooters Supply—www.midsouthshooterssupply.com
MidwayUSA—www.midwayusa.com
Missouri Bullet Company—www.missouribullet.com
MTM Case-Gard—www.mtmcase-gard
Norma Precision—www.norma-usa.com
North Fork Technologies Premium Bullets—www.northforkbullets.com
Hawk Bullets—www.hawkbullets.com
Nosler—www.nosler.com
Oehler Research, Inc.—www.oehler-research.com
Oregon Trail Bullet Company—www.laser-cast.com
Pacific Tool & Gauge—www.pacifictoolandgauge.com
Precision Reloading—www.precisionreloading.com
PowerBond Bullets—www.powerbondbullets.com
Ranier Ballistics Corporation—www.ranierballistics.com
ReloadersAuction.com
Reloading Specialties, Inc.—www.reloadingspecialtiesinc.com
Remington—www.remington.com
Rim Rock Bullets—www.rimrockbullets.net
RockTumbler.com
Shooting Chrony, Inc.—www.shootingchrony.com
Sierra Bullets—www.sierrabullets.com
Speer Bullets—www.speerbullets.com
Sportsman's Warehouse—www.sportsmanswarehouse.com
Starline—www.starlinebrass.com
Swift Bullets—www.swiftbullets.com
Target Sports USA—www.targetsportsusa.com
Thumler's Tumbler—www.thumlerstumbler.com
Top Brass, Inc.—www.topbrass-inc.com
Widener's Shooting Supply—www.wideners.com
Winchester—www.winchester.com
Wolf Performance Ammunition—www.wolfammo.com

Woodleigh Bullets—www.woodleighbullets.com
Zero Bullet Company—www.zerobullets.com

SHOTSHELL PRESSES, DIES, AND TOOLS

Dillon Precision—www.dillonprecision.com
Hornady—www.hornady.com
Lee Precision—www.leeprecision.com
Mayville Engineering Company (MEC R
Ponsness-Warrens
RCBS—www.rcbs.com
Spolar—www.spolargold.com

SHOTSHELL SHOT HULLS, PRIMERS, WADS AND RELOADING ACCESSORIES

Ballistic Products, Inc.—www.ballisticproducts.com
Brecknell—www.brecknellscales.com
Brownell's—www.brownells.com
Buffalo Arms Co.—www.buffaloarms.com
Cabela's—www.cabelas.com
Claybuster Wads—www.claybusterwads.com
Cheaper Than Dirt—www.cheaperthandirt.com
Circle Fly Wads, LLC—www.circlefly.com
Downrange Manufacturing—www.downrangemfg.com
Federal Premium—www.federalpremium.com
Gamaliel Shooting Supply—www.gamaliel.com
Graf & Sons—www.grafs.com
Lohman Arms—www.lohmanarms.com
Mayco Industries, Inc.—www.maycoindustries.com
MAXAM Outdoors—www.maxamcomponents.com
MidSouth Shooters Supply—www.midsouthshooterssupply.com
MidwayUSA—www.midwayusa.com
MTM Case-Gard—www.mtmcase-gard
Natchez Shooter's Supply—www.natchezss.com
Oehler Research, Inc.—www.oehler-research.com
Pacific Tool & Gauge—www.pacifictoolandgauge.com
Precision Reloading—www.precisionreloading.com
Quality Cartridge—www.qual-cart.com
Remington—www.remington.com
RotoMetals—www.rotometals.com
Santa Rosa Lead Products—www.santarosaleadproducts.com
Shooting Chrony, Inc.—www.shootingchrony.com
Sinclair International—www.sinclairintl.com
SlugsRUs—www.slugsrus.com
Target Sports USA—www.targetsportsusa.com
Widener's Shooting Supply—www.wideners.com
Winchester—www.winchester.com
Zipmetals—www.zipmetals.com

POWDER

Accurate—www.accuratepowder.com
Alliant—www.alliantpowder.com
Hodgdon—www.hodgdon.com
IMR—www.imrpowder.com
MAXAM Outdoors—www.maxamcomponents.com
Ramshot—www.ramshot.com
Vihtavouri—www.vihtavouri-lapua.com
Western Powders—www.westernpowders.com
Winchester—www.wwpowder.com

RELOADING BOOKS/VIDEOS AND LOAD DATA

Reloading for Handgunners, by Patrick Sweeney—www.gundigeststore.com
Cartridges of the World, 13th Edition—www.gundigeststore.com
Barnes Reloading Manual Number 4—www.barnesbullets.com
Berger Bullets Reloading Manual—www.bergerbullets.com
Hodgdon Reloading Data Center—data.hodgdon.com
Hornady 9th Edition Handbook of Cartridge Reloading—www.hornady.com
Lee Precision Modern Reloading, 2nd Edition—www.leeprecision.com
Lyman 49th Edition Reloading Book—www.lymanproducts.com
Lyman Load Data Series—www.lymanproducts.com
Lyman Shotshell Reloading Book, 5th Edition—www.lymanproducts.com
Joyce Hornady & Metallic Reloading DVD—www.hornady.com
Nosler Reloading Guide 7—www.nosler.com
RCBS Handbook of Shotshell Reloading—www.midwayusa.com
Sierra 5th Edition Rifle and Handgun Reloading Manual—www.sierrabullets.com
Speer Bullets Reloading Manual #14—www.speer-bullets.com
www.brownells.com (select the "Books & Video" tab)
www.midwayusa.com (select "Books, Videos & Software," then "Reloading Books & Videos)

FREE DOWNLOADS—
NO STRINGS ATTACHED.
No, Really!
Hone your expertise on the latest concealed carry guns, AR-15 optics, ammunition options, shooting tactics and more from our collection of more than 20 topic-specific downloads at gundigest.com/free. And hey, the price is right—they're free!
GunDigest
GUIDE TO AMMUNITION
Cartridge Basics Explained
GunDigest
AR-15 OPTICS
Optics Ideas for AR-15 Rifles
GunDigest
CONCEALED CARRY GUNS
New Semi-Auto Pistols for Self-Defense
GunDigest
ESSENTIAL GLOCK ACCESSORIES
Must-Have Glock Parts and Upgrades
GunDigest
HANDGUNS A BEGINNER'S GUIDE
Choose the Best Self-Defense Handgun
MORE DOWNLOADS
BOOKMARK US!
COMING SOON...
Download now at
GunDigest.com